Hiroshima to Fukushima

Science Policy Reports

The series Science Policy Reports presents the endorsed results of important studies in basic and applied areas of science and technology. They include, to give just a few examples: panel reports exploring the practical and economic feasibility of a new technology; R & D studies of development opportunities for particular materials, devices or other inventions; reports by responsible bodies on technology standardization in developing branches of industry.

Sponsored typically by large organizations – government agencies, watchdogs, funding bodies, standards institutes, international consortia – the studies selected for Science Policy Reports will disseminate carefully compiled information, detailed data and in-depth analysis to a wide audience. They will bring out implications of scientific discoveries and technologies in societal, cultural, environmental, political and/or commercial contexts and will enable interested parties to take advantage of new opportunities and exploit on-going development processes to the full.

For other titles published in this series, go to
http://www.springer.com/series/8882

Eiichiro Ochiai

Hiroshima to Fukushima

Biohazards of Radiation

Springer

Eiichiro Ochiai
Vancouver
British Columbia
Canada

ISSN 2213-1965 ISSN 2213-1973 (electronic)
ISBN 978-3-642-38726-5 ISBN 978-3-642-38727-2 (eBook)
DOI 10.1007/978-3-642-38727-2
Springer Heidelberg New York Dordrecht London

Library of Congress Control Number: 2013948873

Printed on acid-free paper

Springer is part of Springer Science+Business Media (www.springer.com)

Preface

A very strong earthquake (of magnitude 9.0) and huge Tsunami (more than 30 m high) hit the northeast coast of Japan's main island on March 11th, 2011 (3.11 disaster). It devastated many coastal cities, towns, and villages, wiping out the lives and livelihood of many people. It also hit a nuclear power plant: Fukushima Dai-ichi (#1) Nuclear Power Plant of Tokyo Electric Power Co (TEPCo). The quake damaged many parts of the plant, cut off the main electricity source, and also caused failure of the cooling systems, which are required to maintain the nuclear fuel rods at a low temperature. The bad situation was exacerbated by the devastating Tsunami, which destroyed the emergency electricity sources. This caused a number of incidents that released radioactive material produced in the nuclear fuel rods. The four reactors (units 1–4) were subjected to various kinds of explosions.

The explosion of the unit 3 reactor reminded us of the atomic bombs dropped on Hiroshima and Nagasaki in 1945. The extent of the explosion and the damage caused are not comparable, but there are a number of issues common to both of them. One of the issues is that they released enormous amounts of radioactive material. In other words, Japan has been hit by radioactive material (fallouts; sometimes called the "ash of death" in Japan) now three times: two due to the atomic bombs and now the accident at Fukushima nuclear power plant.

There was another incident in 1954; the fishermen on Dai-go (#5) Fukuryu-maru were exposed to the fallout produced by a hydrogen bomb test by the United States on Bikini Atoll in the middle of the Pacific Ocean on March 1st, 1954. It has been revealed recently that this incident was much larger in scope and much worse than what was publicly revealed at the time. More than 900 fishing vessels were exposed to the fallout, and they had to dispose of their catch. Two hundred and forty one fishermen were exposed to radiation, and 77 died by May 1988; 61 of those 77 died of cancers. On March 24th, another hydrogen bomb was exploded, and 19 of the fishermen on Dai-ni (#2) Kosei-maru were radiation-exposed and died in their 40s and 50s. The radiation on the boat was 48.5 mSv/h. (And many of the islanders are still suffering from the effects of radiation as well.) If this is included in the counting, Japan has now experienced four (not three) major incidences of radiation exposure.

The effect of radiation due to the Fukushima disaster is not confined to Japan, as was seen in the event of the Chernobyl nuclear power plant accident in 1986. There are currently over 440 nuclear power reactors operating on this planet, but some developing countries such as China and India are planning to build more in the future; they will definitely increase the dangers of radiation.

Radioactive materials have also been dispersed as a result of atomic bomb tests above ground as well as underground. At least 2,000 aboveground tests were conducted, and all the radioactive products associated with the atomic bomb explosions would have fallen all over the world. A least estimate of these fallouts from the tests is said to be equivalent to that of 40,000 Hiroshima bombs. In addition, the nuclear waste from nuclear power reactors was dumped into the ocean until an international ban came into effect in 1993. All of these human activities have spread an enormous amount of radioactive material in the environment in the last half a century or so, and have increased the level of background radiation. Some of the radioactive isotopes have disappeared since, but many of them are still present in the oceans and on land. This could be at least partially responsible for the dramatic increase of cancers in recent decades.

This treatise will deal with the following issues: (1) the overall picture of the nuclear industry (weapons and power reactors), (2) the scientific bases of nuclear reactions versus chemical reactions and the biohazards of radiation, (3) some relevant data obtained from the Hiroshima/Nagasaki atomic bombs, the Chernobyl and Fukushima incidents, and others, and (4) how the science about the radiation effects has been developed, manipulated, and suppressed throughout the history of the development of the nuclear industry.

One of the basic themes of this essay will be to show that "Life is Incompatible with High-Energy Radiations such as α, β, and γ". The second would be to point out the necessity of the independence of science from political and economic constraints for humans to obtain accurate scientific data so that the human race would gain enough wisdom to succeed in surviving this mire of radiation effects.

It must be noted at the outset that not many strictly scientific studies have been carried out on the low-level radiation effects on health, because of the difficulties in obtaining accurate enough data and, also, the nature of the effects, which are essentially stochastic. Causality can be demonstrated only through epidemiological studies, which require large sample cohorts and proper control groups. Neither of these is easy to obtain or are often impossible in the case of low-level radiation effects. Therefore, no strict causality has been well established except for a few cases, and what we have gained so far are mostly correlations between health effects and radiation. Correlation does not necessarily prove causality, unless all other possible correlation factors can be proven wrong. Particularly difficult is the nature of the health effects, as the causes for diseases are known to be multiple and complicated.

Another difficulty needs to be pointed out here. Not many studies have been published in "proper" scientific journals, not because studies have not been conducted, but often because studies of controversial ideas/results have been denied publication. Therefore, much relevant information may need to be cited from rather uncommon and obscure sources, including internet sites. The accuracy of the information given in these sources is often not strictly verifiable, and, hence, we should be cautious of the information thus obtained.

Besides, many important original literatures regarding the most serious issues, i.e., Chernobyl, were published in Russian, which this writer has neither access to nor the ability to understand. He has to rely on English translations or literature written in English about those source materials written in Russian.

The author gratefully acknowledges the useful comments and suggestions on several portions of the manuscript made by Dr. Leonard Angel, Emeritus Professor of Philosophy, Douglas College, Vancouver, Canada. It must be pointed out, though, that the author alone is responsible for all the ideas and interpretations of data expressed in this book, and any mistakes within it.

It might be suggested that readers who are not familiar with or not interested in the scientific aspects of the issues may skip Part I through Part IV.

Acknowledgement

The author wishes to express sincere thanks for the copyright owners for granting him permissions to reuse the copyrighted material in this book, as indicated in the legends of the figures.

Contents

Part I The Nuclear World, the Chemical World, and Radiation

1 The Nuclear World 3
1.1 Atoms 3
1.2 Nucleosynthesis 4
1.3 Nuclear Binding Energy 6
1.4 Nuclear Reactions 7
1.5 Note 8
1.5.1 Quantity = Number of Particles and Mass of Nucleon Particles 8
1.5.2 The Unit of Energy, Its Relationship to Mass, Temperature, and Frequency of the Photon Equivalent 9
1.5.3 Some Physical Constants 10
1.5.4 Scales of Number 10
1.5.5 Periodic Table: Periodic Chart 10
References 12

2 Nuclear Power and Radiation 13
2.1 Nuclear Fission Reaction 13
2.2 Nuclear Chain Reaction and Nuclear Power 14
2.3 Atomic Bombs and Nuclear Reactors 15
2.4 Radioactive Isotopes and Their Decay 16
References 20

3 The Chemical World 21
3.1 The Basics 21
3.2 Nonliving Chemical World 23
3.3 Living Chemical World 24
References 25

Part II The Military Uses of Nuclear Power: Hiroshima

4 Developing Atomic Bombs: The Manhattan Project 29
4.1 Birth of the Atomic Bomb 29
4.2 Why Did the United States Drop Atomic Bombs on Japan? 30
4.3 Nuclear Arms Race 31
References 33

5 Devastation Caused by the Atomic Bombs: Hiroshima and Nagasaki 35
References 37

6 Other Nuclear Weapons 39

Part III The "Peaceful" Uses of Nuclear Power: Fukushima

7 The Development of Nuclear Power Reactors and the Nuclear Industry 43
7.1 General Development 43
7.2 Nuclear Power Industry in Japan 45
7.3 General Problems 47
Reference 48

8 Major Incidences of the Failure of Nuclear Facilities 49
8.1 Radiation Dispersal Under Normal Conditions 49
8.2 A Brief History of Major Incidents of Radiation Release 51
8.3 The Accident at Three Mile Island Nuclear Power Plant 52
8.4 The Accident at Chernobyl Nuclear Power Plant 53
8.5 The Accident at Fukushima Dai-ichi Nuclear Power Plant 56
8.5.1 The Causes and Processes of the Accident 56
8.5.2 The Distribution of Radioactive Material Released 60
References 67

Part IV The Interactions of Radiation with the Chemical/Biological World: The Biohazards of Radiation

9 Interactions of Radiation with the Chemical World 71
9.1 Energy Changes in the Chemical World 71
9.2 Interaction of High-Energy Radiation with Chemicals 73
9.2.1 Photons (X-rays and γ-rays) 73
9.2.2 Beta (β)-Particles (Electrons) and β^+-Particles (Positrons) . . . 75
9.2.3 Alpha (α)-Particles 75
9.2.4 Neutrons 76
9.2.5 Radiation on Nonliving Chemicals 76
Reference 78

10 Chemicals and Their Behaviors in Biological Systems 79
10.1 A Survey 79
10.2 Potassium and Cesium 80
10.3 Calcium and Strontium 84
10.4 Other Elements 85
References 86

11 The Effects of Radiation on Biological Systems 87
11.1 Expression of Dose Due to Exposure: Sievert 87
11.1.1 Radioactivity and Exposure Dose Expression Currently in Use 87
11.1.2 Ambiguity in Exposure Dose Expression: Sv (Gy) 88
11.1.3 Exposure Effects, an Alternative Expression 89
11.2 External and Internal Exposure 90
11.3 Radioactive Material, Natural and Artificial, and Background Exposure 92
11.3.1 Radiation Sources 92
11.3.2 Background Radiation 93
11.3.3 Artificial Sources of Radiation: Medical Uses 95
11.3.4 Atomic Bombs 96
11.3.5 Radiation Escaped from Nuclear Power Reactors 97
11.4 Acute Radiation Effects 97
11.4.1 General Trends 97
11.4.2 Atomic Bomb Victims 100
11.4.3 Other Incidents 101
11.4.4 Acute Internal Exposure Effects: Atomic Bombs 102
11.5 Low-Dose Internal Exposure Effects 102
11.5.1 How and in What Form the Radiation Particles Were Spread from the Explosion of an Atomic Bomb? 103
11.5.2 Release and Spread of Radioactive Material from a Nuclear Power Plant Due to an Accident 104
11.5.3 How Radioactive Material (Gas or Small Particles) May Enter the Human Body and How They Behave? 105
11.5.4 Basic Mechanisms of Internal Exposure Effects 106
11.5.5 Stochastic Nature of Low-Dose Radiation Effects 109
11.5.6 Issue of LNT and Acute Versus Protracted Exposure 110
References 111

12 Control of Cell Processes and Defenses Against Radiation 113
12.1 Mechanisms to Defend Against Damaging Effects or to Repair Lesions on Non-DNA Molecules 113
12.1.1 Effects of Nonradioactive Entities 113
12.1.2 Remedies for Radiation Effects 115

12.2 Control of Cellular Processes 118
12.2.1 Organism/Organs/Tissues/Cells 118
12.2.2 Cell Cycle 119
12.3 Damage to DNA and Controlling Systems 121
12.3.1 Damage/Modifications Caused by Nonradioactive Chemicals 122
12.3.2 Lesions on DNA Caused by Radiation 124
12.3.3 Cell Cycle Control upon DNA Damage 126
12.3.4 Repair of Damaged DNA 128
12.4 Cancer-Causing Processes 131
12.4.1 Processes Leading to Cancerous States 131
12.4.2 Specific Effects by Radiation? 134
12.5 Are Defense Mechanisms Sufficient? 136
12.6 Transgenerational Transmission of Radiation Effects? 138
References 139

Part V Health Effects of Radiation Revealed So Far

13 Data on Hiroshima and Nagasaki 143
13.1 Acute External Exposure Effects 143
13.2 Hiroshima/Nagasaki: Long-Term Effects 146
13.2.1 Slow-Manifesting Cancers: General Trend 146
13.2.2 Radiation-Related Leukemia 147
13.2.3 LSS-14 Report 147
13.2.4 Problems with the Analysis of LSS Reports 150
13.3 Effects of Fallouts on Other Parts of Japan 151
References 152

14 Data on Chernobyl 155
14.1 Discrepancy Between the Authorities' Reports and the Reality . . . 155
14.2 Liquidators 156
14.3 Cs-137 Internal Effects on Noncancerous Diseases 157
14.3.1 The Overall Distribution 157
14.3.2 Internal Exposure 159
14.4 Cancers 163
14.4.1 Thyroid Cancer 163
14.4.2 Other Cancers 164
14.5 Other Effects 165
14.5.1 Infant Mortality 165
14.5.2 Teratogenic Effects (Malformations) and Others 166
14.5.3 Effects on the Brain: Psychological and Mental Problems 167
14.6 Transgenerational Effects 168
14.7 Effects on Plants and Animals 168
14.7.1 Plants 168
14.7.2 Animals and Microorganisms 169
References 171

15 Data on Fukushima 173
15.1 Infant Mortality 173
15.2 Heart Failure and Other Diseases 173
15.3 Thyroid Gland Abnormalities and Cancers 176
15.4 Other Indications 177
15.5 Current Situations 178
References 178

16 A-Bomb Tests; Three Mile Island and Other Incidents; NPPs Under Normal Operation; Depleted Uranium Bombs 179
16.1 Effects of A-Bomb/H-Bomb Tests 179
16.2 Three Mile Island Incident 181
16.3 Effects from NPPs and Other Nuclear Facilities Under "Normal" Operational Conditions 183
16.3.1 In Europe 183
16.3.2 In the United States 184
16.4 DU (Depleted Uranium) Munitions 186
16.4.1 What Are DU Munitions and Where Have They Been Used? 186
16.4.2 Gulf War Syndrome Among Veterans 188
16.4.3 DU Effects in Yugoslavia (Bosnia, Herzegovina, Serbia) 189
16.4.4 Iraq/Fallujah/Basra 189
16.5 An Overall Trend in Cancer Rates 191
References 194

17 (High-Energy) Radiation Is Incompatible with Life 197
17.1 Incompatibility of Radiation with Life 197
17.2 Radiation in the Evolution of Life 200
17.3 Difficulties in Combating Radiation Effects in Real Life 201
References 203

Part VI The Science of Biohazards of Radiation Has Been Influenced by Non-scientific Interests

18 Falsification and Suppression of the Truth 207
18.1 Development of the Idea of "Radiation Biology" 207
18.2 Manhattan Project up to Hiroshima/Nagasaki and Subsequent Developments 208
18.3 Atomic Bomb Tests 210
18.4 Three Mile Island Incident 211
18.5 Chernobyl Incident 212
18.6 Fukushima Incident 213
References 215

19 The Crisis of Human Civilization and Life-Supporting Planet Earth 217
19.1 Nuclear Industry Lobby and Their Tactics 217
19.2 Other Signs of Radiation Effects? 218
19.3 Beyond Radiation Issues 220
References 222

Postscript 223

Part I
The Nuclear World, the Chemical World, and Radiation

The whole universe, as it is currently understood, consists of four forces: strong and weak forces, which govern the behaviors of fundamental particles, quarks, nucleons, and their transformations; electromagnetic force, which governs the chemical world, though operating in the nucleus as well; and gravity. Radiation is associated with nuclear transformation, and, hence, is governed by the strong and weak forces, which are typically several thousands to millions times as large as the typical electromagnetic forces acting in the chemical world. This is the fundamental problem in the radiation effects on life. This part deals with the fundamentals of the nuclear world, nuclear power, the chemical world, and radiation.

Chapter 1
The Nuclear World

A reader who is not familiar with some scientific notations, units of quantity, and energy, as well as the periodic table (a list of all the elements in the universe) might refer to Sect. 1.5 at the end of the chapter before starting to read this first chapter.

1.1 Atoms

The issues we will be dealing with in this treatment are nuclear reactions, such as fission reactions, radiation from an unstable nucleus, and the interaction of radiation with the chemical world, of which we, living organisms, are a part. The chemical world consists of atoms and their aggregates, molecules (chemical compounds in general). As far as we know, the material world encompasses the entire universe, and is made up of about 100 or so elements in the chemical sense. The material is composed of atoms, and a single atom consists of electrons and a nucleus, which is made of nucleons: protons and neutrons. It is schematically shown in Fig. 1.1. In reality, the nucleus is a very small positively charged blob of size on the order of magnitude of 10^{-15} to 10^{-14} m, whereas the electrons look more like a negatively charged cloud, the size of which is on the order of 10^{-10} m, and this represents the size of an atom. Atoms combine to form chemical compounds, which constitute the material world, including the earth and the living organisms on it. This is a brief description of the material world, i.e., the universe. We will look at it and see its implication for our concern here, i.e., the radiation effects on living organisms.

The "chemical" behaviors of atoms are determined mostly by the electric charge Z of the nucleus and the number of electrons surrounding it. The total number of nucleons, i.e., protons Z and neutrons N, in the nucleus of an atom is defined as the "mass number" (A of the nuclide = Z + N), and the number of protons, i.e., the electric charge, is called the "atomic number" (Z or nuclear charge) of the atom. In the chemical world, an atom has several electrons around the nucleus. A specific atom with A and Z is expressed as ${}_{Z}X^{A}$, where X is the symbol of the element. Atoms with the same Z but different A belong to the same element X, and there is

E. Ochiai, *Hiroshima to Fukushima*, Science Policy Reports,
DOI 10.1007/978-3-642-38727-2_1, © Springer-Verlag Berlin Heidelberg 2014

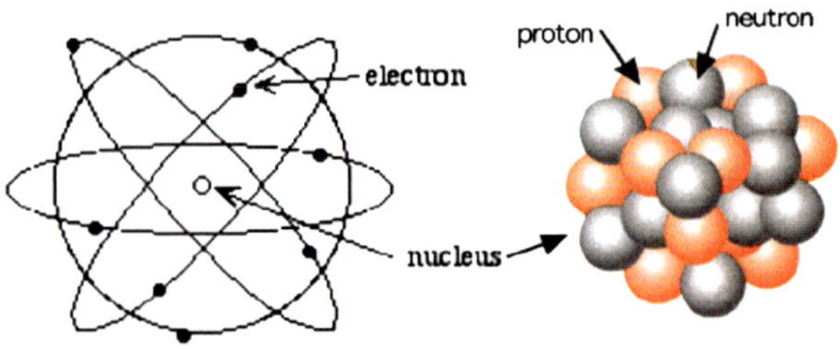

Fig. 1.1 Structure of the atom

usually one to several tens of different nuclides which have the same Z but different A; these are called "isotopes". Isotopes of a Z value behave chemically in the same manner, i.e., independent of the N or A values. However, the nuclear behaviors among isotopes of the same Z value differ widely depending on the N value. Some are stable, but others are not and, hence, disintegrate spontaneously.

1.2 Nucleosynthesis

Everything we know is supposed to have originated from the "Big Bang". There was something enormously dense and of enormously high temperature, which exploded about 13.7 billion years ago. At that point, the energy of the universe (made of photons) is believed to be something like 10^{19} GeV, and the three major forces (the strong nuclear force, weak nuclear force, and electromagnetic force) were still unified.

At about 10^{-35} s after the Big Bang, the inflation of the universe became accelerated, and baryons such as quarks that are the basis for the eventual creation of nucleons started to form, as the temperature became lower (*ca.* 10^{15} GeV), though it was high enough to not yet allow the formation of other kinds of particles. At that point, the "strong force" started to diverge from the grand unified status of forces. The formation of matter is the conversion of photons to matter and antimatter; they should have formed in equal quantity, but they soon annihilated each other, turning back into photons. Somehow, there was asymmetry, which allowed a slightly greater abundance of matter remaining over the antimatter. Hence, this has eventually led to today's universe where "matter" dominates.

At about 10^{-11} s (*ca.* 200 GeV temperature wise), the two forces "weak force" and "electromagnetic force" started to be separated, and, hence, the present three forces (plus gravity) appeared separated.

At about 10^{-5} s, baryons started to combine to form nucleons, protons, and neutrons (through the "strong force"). Light particles called leptons also appeared; leptons include electrons, muons, and neutrinos. At this high temperature, electrons still do not combine with protons (to form an atom). A neutron first combines with a proton to form deuterium (heavy hydrogen nucleus); this nucleosynthesis started about 3 min after the Big Bang. The residual "strong force", called "nuclear force", and electromagnetic force govern the nucleosynthesis. The "residual" part implies that the strong force is mostly used to form nucleons from quarks, and the remaining portion of the strong force is used to bind nucleons to form a nucleus. This nuclear force is still much stronger (by several orders of magnitude) than the electromagnetic force. Soon, deuterium binds another neutron to form tritium, which then absorbs a proton to form the helium nucleus. The last process requires a very high temperature, as a positively charged proton has to collide with another positively charged tritium nucleus; this easily happened in the early universe due to the high temperature (see, for example, Silk (1997)).

Today, helium is being synthesized by the thermonuclear fusion reaction inside stars like the Sun. The temperature at the core of the Sun is about ten million degrees, and hydrogen gas is packed at a density of about 81 g/cm^3. Under this condition, hydrogen burning takes place, resulting in the formation of helium nuclei from four protons each (hydrogen nuclei). The energy released from this fusion reaction is 28 MeV. As the temperature in the core increases to 100 million degrees (due to further compression and the heat produced by the hydrogen burning) and the density to 6×10^4 g/cm^3, the star becomes a "red giant", and helium burning commences, resulting in the formation of carbon nucleus ($_6C^{12}$). At higher temperatures, other nucleosyntheses take place. Carbon burning results in many nuclei; oxygen nucleus $_8O^{16}$ from $_6C^{12}$ and helium nucleus ($_2He^4$ α-particle), with the release of γ-rays during the process. $_8O^{16}$ reacts further with an α-particle, resulting in $_{10}Ne^{20}$ (+ γ); two $_6C^{12}$ combine to form $_{12}Mg^{24}$ (+ γ) and $_{11}Na^{23}$ (+ proton). At still higher temperatures, $_8O^{16}$ burning takes place, forming many other heavier nuclei.

Neutrons carry no electric charge, and, hence, would encounter no repulsive force from the entering nucleus. A neutron is added, resulting in a nucleus with one mass number higher, which may be an excited state releasing γ-particles or β-decaying. The latter will result in an isotope of an element which is one value higher in the atomic number. In this way, neutron addition has created many heavier isotopes up to Bismuth ($_{83}Bi^{209}$). Nuclei (isotopes) heavier than this Bi nucleus decay by α-emission. They are created in supernovae (for an accessible account, see Cohen (1967)).

All isotopes have been, and are being, created in the universe where extremely high temperature and pressure prevail, so that nucleosynthesis can take place; hence, the stars, including our own Sun, can be said to be "thermonuclear reactors". Since shortly after its formation such extreme conditions were not prevalent on the earth, thus, nucleosynthesis has not occurred on our planet. All kinds of isotopes (nuclei), stable or unstable, existed at the beginning, and stable isotopes have persisted without changes. These stable isotopes are the basis for the material world of the earth: the core, rocks, rivers, water, and living organisms, including plants, animals, and today's human race. This is the "chemical world". The majority of unstable isotopes have disappeared since, as their half-lives are short, and only the radioisotopes with very long half-lives are still present on the earth. Examples of this last category include $_{92}U^{235}$, $_{92}U^{238}$, $_{90}Th^{232}$, and $_{19}K^{40}$. Some radioisotopes such as $_{88}Ra^{226}$ and $_{86}Rn^{222}$ are found on today's earth, despite their relatively short half-lives, because they are being formed as products of the decaying process of long-lasting U-238 and others.

1.3 Nuclear Binding Energy

The force that holds protons and neutrons together in a nucleus is the "strong nuclear force", which acts over a very short distance, the size of nucleus, i.e., on the order of 1–10 fm. The nuclear state is quantized, as the electronic state in an atom is quantized. The state that protons and neutrons take is defined as an orbit, as in the electronic state, and is specified by quantum numbers, n, l, j, and m, though these quantum numbers are not exactly the same as those in the electronic states. The principal quantum number n takes the values 1, 2, 3,…., and $l = n-1$, $n-3$,…., 1, or 0 (this is different from the l quantum number in the electronic state). No detailed description of the nuclear state is intended here. Only the fact that nucleons would occupy quantized orbits should be kept in mind in seeing nuclear reactions (for an accessible account, see Cohen (1967)).

It requires an enormous amount of force and energy to separate nucleons in a nucleus into individual nucleons. This energy has been measured for many nuclei, and the average value per nucleon is called the "nuclear binding energy". Figure 1.2 shows the nuclear binding energy plotted against the mass number (of the nucleus). There are several peaks; at ${}_2He^4$, ${}_6C^{12}$, ${}_8O^{16}$, and the highest at ${}_{26}Fe^{56}$. Beyond Fe^{56}, the binding energy decreases steadily with increasing mass number or increasing proton numbers. This is caused by the increasing repulsive effect among protons in the nucleus.

It requires 2 MeV to separate a proton and a neutron in a deuterium nucleus, but $7 \times 4 = 28$ MeV to separate two protons and two neutrons in He completely. The two protons and two neutrons fill the orbits of $n = 1$ in the lowest state of an He nucleus (α-particle), and behaves like an inert gas in the sense that the state is in a fully occupied shell defined by n and is very stable, i.e., it has a high binding energy. By the way, the chemical element He (α-particle surrounded by two electrons) happens to have a completed electronic shell (n = 1 is filled) and be chemically inactive. The binding energies of ${}_6C^{12}$, ${}_8O^{16}$, and ${}_{26}Fe^{56}$ are about 7.8, 8, and 8.7 MeV, respectively. In contrast, for example, ${}_7N^{16}$ has the state where 7 protons occupy $n = 1$ and 2 shells (with one vacancy), and 8 out of 9 neutrons complete $n = 1$ and 2 shells, with a single neutron in a much higher $n = 3$ shell. Hence, it is very unstable, and will either emit a neutron or the 9th neutron changes into a proton, emitting an electron (β) and an antineutrino. The result is the decaying of ${}_7N^{16}$ into ${}_8O^{16}$ by emitting β-particles.

The nuclear binding energies are enormously large as compared to a typical chemical energy. A basic chemical binding energy may be represented by the energy to separate the electron from a hydrogen nucleus (to ionize the hydrogen atom). The average distance of the electron in the lowest energy state ($n = 1$) from the nucleus (proton) is 52.9 pm, and the energy required to move the electron far

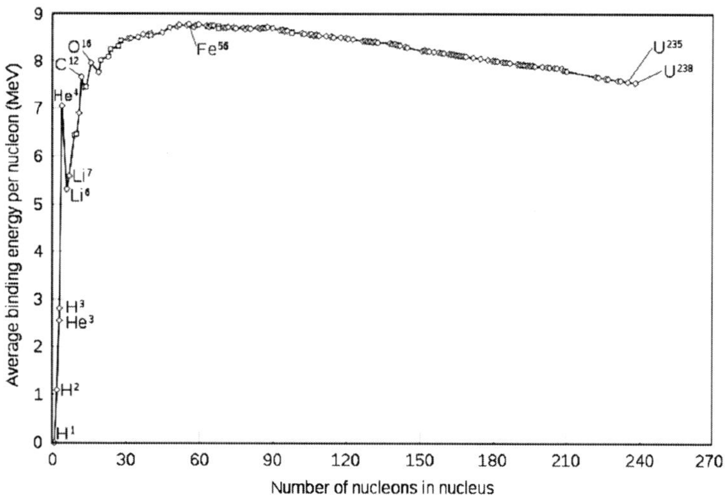

Fig. 1.2 The nuclear binding energy (from Wikipedia)

away can be calculated to be 13.6 eV. This is the electrostatic interaction. Compared to this basic energy of the electron–nucleus interaction, i.e., the basis of chemical energy, the binding interaction energy of nucleons in a nucleus is enormously large, being several million times greater. This difference in binding energy is reflected in the biological effects of radiation discussed later.

1.4 Nuclear Reactions

All kinds of nuclear reactions are possible, but they usually involve a huge amount of energy and would not take place under ordinary (earthly) conditions, except for one type (nuclear decay reactions discussed in the next chapter). The only relatively easy reaction is that between a neutron and a nucleus. As a neutron carries no electric charge, it can approach a nucleus relatively easily, as no electrostatic repulsion opposes it. This is the basis of nuclear fission reactions.

However, the discovery of neutrons was relatively recent (1932), and the earlier studies of nuclear reactions were based on the bombardment of high-energy α-particles from known α-emitters, such as U-238. The first ever artificial nuclear reaction was conducted by E. Rutherford in 1919. It was later confirmed by P. M. S. Blackett that the reaction can be written as:

$$ {}_2\mathrm{He}^4(\alpha) + {}_7\mathrm{N}^{14} \rightarrow {}_8\mathrm{O}^{17} + {}_1\mathrm{p}^1\ (\text{proton}) $$

In 1937, one of the missing elements from the periodic table, the atomic number 43, was created by bombarding molybdenum (atomic number 42) with neutrons

(E. Segre). This element was nonexistent on the earth, but is created by humans and is called “technetium” (Tc). The reaction of its formation can be written as:

$$_{42}Mo^{96} + {_0}n^1(\text{neutron}) \rightarrow {_{43}}Tc^{97} + {_{-1}}e^0(\text{electron} = \beta\text{-particle})$$

The so-called transuranium elements (atomic number ≥93) are created by scientists. One important transuranium element that is produced in nuclear fuel rods and atomic bombs is $_{94}Pu^{239}$ (240 and 241 as well), which forms as a result of neutron capture by U-238. It forms U-239, which rapidly turns into $_{93}Np^{239}$ by β-emission, which then turns into Pu-239, again through β-emission, as shown below (T is the half-life);

$$_{92}U^{238} + {_0}n^1 \rightarrow {_{92}}U^{239} - (T = 23.5\ \text{min}, {_{-1}}e^0\beta\text{-radiation}) \rightarrow$$
$$_{93}Np^{239} - (T = 2.4\ \text{days}, {_{-1}}e^0\ \beta\text{-radiation}) \rightarrow {_{94}}Pu^{239}$$

An unstable nucleus undergoes several kinds of nuclear reactions to attain a more stable form. This change is called nuclear “decay” (or disintegration). The decay process is spontaneous and proceeds at a certain speed (depending on the nucleus). The decay process will inevitably emit high energy in the form of particles, α, β, and γ (photons). This topic will be covered in the next chapter.

1.5 Note

The expressions of quantity and energy used in this treatment will be rather exotic, and need to be understood in order to see this entire issue which encompasses nucleons, nuclei, atoms, electrons, chemical compounds, and the human body, the Sun, and the universe. A list of elements, i.e., the periodic table, is shown below.

1.5.1 *Quantity = Number of Particles and Mass of Nucleon Particles*

Material is composed of minute particles. A typical material, water, consists of particles of a molecule; a water molecule consists of two hydrogen atoms and one oxygen atom. An ordinary measure of quantity is mass (weight) in terms of kg (or g) in everyday life. However, a more fundamental measure of quantity is the number of particles. In the case of water, for example, 18 g of water consists of 6.02214×10^{23} particles of water molecules. This number, 6.02214×10^{23}, is called Avogadro’s number and the quantity of a substance consisting of this many molecules is called one “mole”. “Mole” is the unit which is appropriate for discussing the behaviors of atoms, molecules, etc. in everyday life (the chemical world). You can then guess that a single particle of water molecule weighs a miniscule amount, approximately

$18/6.02214 \times 10^{23} = ca.\ 3 \times 10^{-23}$ g ($=3 \times 10^{-26}$ kg). Avogadro's number has been defined in such a way that one mole of the most abundant isotope of carbon (carbon-12; consisting of 6 electrons and a nucleus of 6 protons and 6 neutrons) weighs exactly 12.0000 g. So, 1/12th of a single carbon-12 is used as a unit of atomic mass (corresponding roughly to the mass of a single proton or neutron); this is 1.66054×10^{-27} kg (defined as "u" atomic mass unit). However, the mass of an independent proton or neutron is slightly different from it; m_p (mass of a proton) $= 1.67262 \times 10^{-27}$ kg and m_n (mass of a neutron) $= 1.67493 \times 10^{-27}$ kg.

It must be noted that we will be using weight units of kg or g in everyday life to express quantity, but that we need to use the number of particles (typically a very large number) and its individual mass (typically a very small number) when dealing with the nuclear world.

1.5.2 The Unit of Energy, Its Relationship to Mass, Temperature, and Frequency of the Photon Equivalent

It is necessary to understand the unit of energy and its relationship with temperature and mass, as well as the frequency of its photon equivalent. Therefore, a short description is given here, but the reader may ignore this note if they are familiar with the definitions. The ultimate unit of electric charge is represented by the electric charge of an electron (negative) or that of a proton (positive). Their electric charges are of the same magnitude: 1.60218×10^{-19} C (coulomb). When one such a charge moves between 1 V (volt), the energy change associated is defined as 1 eV (electronvolt). This is the unit of energy often used in the discussion of the behavior of nuclei and nucleons (protons and neutrons). This applies to a single particle, atom, nucleon, etc. This is defined, as seen here, in terms of electromagnetic force/energy, but the same unit is used to express the nuclear energy that is caused by the strong nuclear force, rather than the electromagnetic force.

The electromagnetic force is the force acting between electric charges and magnetic poles, and is effective over a long distance. The strong nuclear force holds positively charged protons and electrically neutral neutrons within a small space (nucleus), and is effective only at a short distance, 10^{-14} m and shorter.

Since C*V = J (joule), 1 eV $= 1.6021 \times 10^{-19}$ J. (Mechanically, energy J is kgm^2s^{-2}.) J is the energy unit for ordinary phenomenon, such as the burning of coal, electricity, etc., in everyday life. The unit calorie used to be used for that purpose, but now it is joule, and 1 cal = 4.184 J. Therefore, it should be kept in mind that the energy value used in discussing the nuclear world is in the unit of eV, which applies to a single particle, but that the energy value in the chemical world is expressed in terms of J, which applies to a collection of an extremely large number of particles.

In the nuclear world, energy is regarded as being equivalent to mass through Einstein's famous equation, $E = mc^2$ (m = mass, c = speed of light). A high-speed particle in the nuclear world may also behave like a wave with energy of

$E = h\nu$ (h = Planck's constant, ν = frequency). Temperature reflects the energy state of an ensemble of a large number of particles, but it can be regarded as representing a temperature-equivalent of a single particle, $E = kT$ (k = Boltzmann constant, T = temperature). All these equations are, therefore, interrelated or equivalent. Let's take 1 eV ($=1.60218 \times 10^{-19}$ J) as energy. It is equivalent to a mass of 1.782×10^{-36} kg and to a temperature of 1.1604×10^{4} degree. If it is regarded as a wave, it has a frequency of 2.418×10^{14} s^{-1} (Hz), which is in the range of the longest visible to infrared light.

1.5.3 *Some Physical Constants*

Some relevant physical constants are listed below:

Mass of a proton (m_p) at rest $= 1.67262 \times 10^{-27}$ kg
Mass of a neutron (m_n) at rest $= 1.67493 \times 10^{-27}$ kg
Mass of an electron (m_e) at rest $= 9.10939 \times 10^{-31}$ kg
Atomic mass unit u $= 1.66054 \times 10^{-27}$ kg
Speed of light c $= 2.99792 \times 10^{8}$ m/s
Electric charge of an electron (or proton) $= 1.60218 \times 10^{-19}$ C
Planck's constant $= 6.62608 \times 10^{-34}$ Js
Boltzmann's constant $= 1.38066 \times 10^{-23}$ J/K
Avogadro's number $= 6.02214 \times 10^{23}$ particles/mol

1.5.4 *Scales of Number*

An ordinary small number such as 785 is expressed as such, but ten times its value, 7,850, is expressed as seven thousand eight hundred and fifty (in English), which can be expressed as 7.85×10^{3} in scientific notation, and also 7.85 kilo (K). At a thousand times this number, it would be 7,850,000 or 7.85×10^{6}, 7.85 million, and 7.85 mega (M). Beyond that, 10^{9} is billion, giga (G); 10^{12} trillion, tera; 10^{15} quadrillion, peta; 10^{18} quintillion, exa; 10^{21} sextillion, zetta.

On the other hand, a much smaller number than one hundredth is as follows: 1/100 is 10^{-2} centi, 10^{-3} milli, 10^{-6} micro, 10^{-9} nano, 10^{-12} pico, 10^{-15} femto, 10^{-18} atto, and 10^{-21} zepto.

1.5.5 *Periodic Table: Periodic Chart*

This is a list of elements systematically arranged (Fig. 1.3). It shows a number of things regarding the chemical characters of the element, as the system is based on the electronic structure of an atom (an element). An element is defined in terms of the positive charge (number of protons) of the nucleus, as it determines the chemical behaviors. Elements in the same column, generally speaking, behave similarly, but

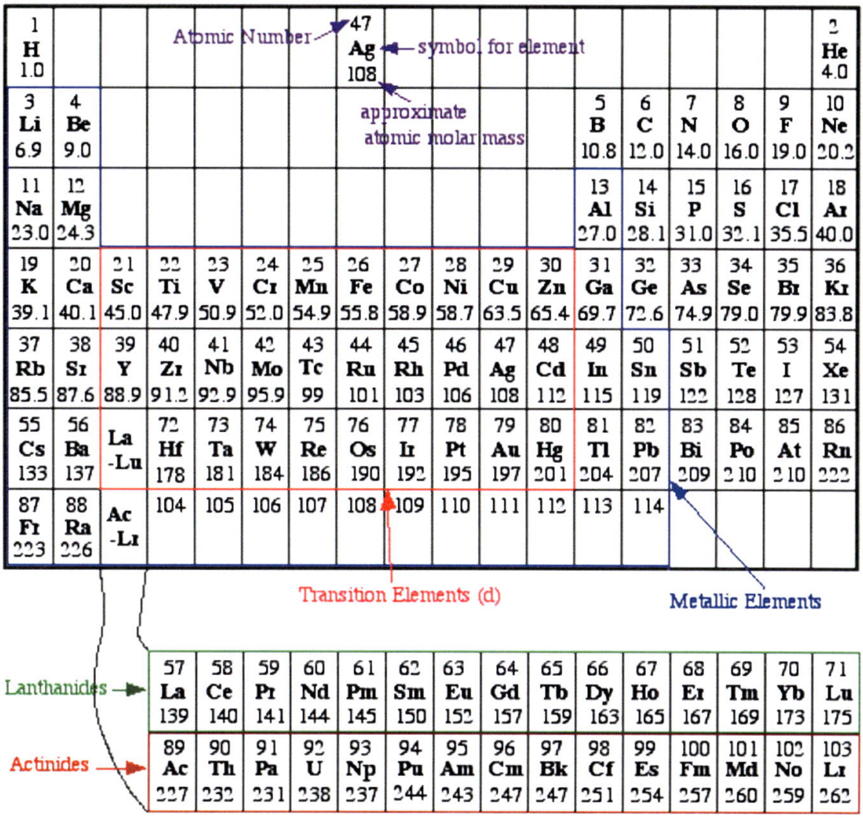

Fig. 1.3 Periodic table of elements

not exactly. The elements in the last column (He, Ne,...) have their electronic structure in a closed-shell form, which is stable, chemically nonreactive, and do not make compounds, except for Xe. They behave as single atoms and are called "inert gas" elements. The elements belonging to the first column are termed as "alkali" metals, and they have a single electron in the outermost electronic shell, and, hence, easily lose the single electron to become +1 ion (Li^+ or Li(+1), Na^+, K^+, Rb^+, Cs^+, etc.). Likewise, the elements in the second column (alkaline earth metals) tend to easily become +2 ion (Be^{2+} or Be(+2), Mg^{2+}, Ca^{+2}, Sr^{+2}, etc.). Typical compounds of the elements starting with B are B_2O_3, Al_2O_3, etc. The next column contains C, Si, Ge, etc., and the typical compounds are CO_2, SiO_2, etc. The behaviors of elements in the next column starting with N are rather complicated. N takes the forms NH_3 or NO_3^- and anything in between. But P and As most often take the forms PO_4^{3-} and AsO_4^{3-}. The elements that start with O usually become O^{2-} or O (−2), S^{2-}, Se^{2-}, etc., but S, Se, and Te easily form ions SO_4^{2-}, SeO_4^{2-}, etc. The elements in the second from last column, on the other hand, tend to grab a single electron, forming the stable electronic structure of the corresponding inert gas. Therefore, F, Cl, Br, and I tend to gain an electron and become F^- or F(−1), Cl^-, Br^-, and I^-, respectively. There are ten columns in between, which contain familiar metallic elements, such as Fe (iron), Co (cobalt), Cu (copper), Ag (silver), and Au (gold). These are called transition (d-)elements. Their behaviors are variable. In addition, there are two sets of 15 elements, each placed outside the main table, but they are contiguous with the main chart, as indicated by their initial element and the numbering. These are f-elements, and the elements in the upper row are called "lanthanides", while the elements in the lower row are called "actinides". The last element in each group Lu, Lr does not belong to the respective group, but is placed there for convenient sake.

References

Cohen BL (1967) The heart of the atom. Doubleday, Garden City
Silk J (1997) A short history of the universe. Scientific American Library, New York

Chapter 2
Nuclear Power and Radiation

2.1 Nuclear Fission Reaction

How the nuclear fission of uranium was discovered is detailed in, e.g., Jeremy Bernstein's "Plutonium" (Bernstein 2007). The phenomenon of nuclear fission itself was first discovered by E. Walton and J. Cockcroft in 1932 with regard to the reaction of proton with $_3Li^7$, which produced two α-particles (helium nuclei). Enrico Fermi and his associates in Rome studied the reaction of neutrons with uranium in 1934. Lise Meitner, along with Otto Hahn and Fritz Strassmann, worked on this issue as well. They found that one of the products was barium. Hahn and Strassmann published their results in December 1938. Hahn was skeptical, but Meitner realized that uranium was split into smaller nuclei, one of which was barium. Hence, she was the first to recognize a nuclear fission reaction (of uranium), but, unfortunately, she did not share the Nobel Prize, which was awarded to Otto Hahn. Meitner was the first Austrian woman physicist and a Jew, and had a lot of trouble during the 1930s because of the rise of the Nazi movement.

A neutron can collide easily with a nucleus, because of its electrical neutrality. However, a neutron with a high speed may often fly by a nucleus without getting captured. Whether it happens that way or not is dependent on the cross section of a nucleus. A nucleus that has a relatively small cross section would be able to capture more easily a slower (termed "thermal") neutron. It turned out that this is the case with uranium ($_{92}U^{235}$; often abbreviated to U-235). The major isotope $_{92}U^{238}$ can be split, but only at a very low speed.

When a neutron enters a nucleus of U-235, it will deform and destabilize the shape of the liquid drop-like nucleus (the Bohr model) and soon split into two smaller nuclei. Once these two positively charged nuclei separate beyond a short critical distance, the strong nuclear force ceases to operate, and electric repulsion pushes them farther apart. This is a fission reaction (Fig. 2.1). A typical fission reaction of U-235 is exemplified by the following equation:

E. Ochiai, *Hiroshima to Fukushima*, Science Policy Reports,
DOI 10.1007/978-3-642-38727-2_2, © Springer-Verlag Berlin Heidelberg 2014

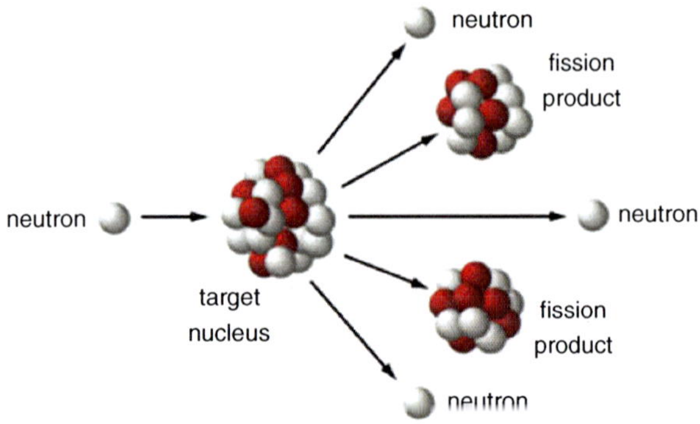

Fig. 2.1 A model of a nuclear fission reaction

$$_{92}U^{235} + {}_{0}n^{1} \rightarrow {}_{56}Ba^{142} + {}_{36}Kr^{92} + 2{}_{0}n^{1}$$

Usually, it would not split into halves exactly, and would produce as many as 200 daughter nuclei in a nuclear reactor reaction, as well as in an atomic explosion. The distribution curve of daughter nuclei as a function of mass number has already been determined. According to this, one daughter nucleus has a mass number peaking at about 135 and the other peaks at about 95. The majority of these fission products is unstable and radioactively decays (discussed below).

We can estimate roughly the energy change in the fission reaction based on the nuclear binding energy (BE) (Fig. 1.2). The BE of U-235 is about 7.7 MeV, and that of the two daughter nuclei are about 8.45 MeV (assuming average mass number = 117). Therefore, the energy change is estimated to be $7.7 \times 235 - 8.4 \times 234 = -168$ MeV. This is the energy change for a single nucleus of U-235 upon fission (the energy carried by an extra neutron is ignored). This energy can be converted to 6.9×10^7 kJ for 1 g of U-235. The energy released is reflected in the loss of mass in the process. In the case of the reaction cited above, the mass loss has been determined to be −0.19 u (see Note 1.5.4 for "u" at the end of Chap. 1). This mass (lost) is converted to energy through $E = mc^2$. This energy can be calculated to be 171 MeV or 7.0×10^7 kJ for 1 g of U-235. This is the basis for the atomic bomb dropped on Hiroshima and the majority of today's nuclear power plants.

Other usable fissionable isotopes are $_{92}U^{233}$ and $_{94}Pu^{239}$. Indeed, the second atomic bomb dropped on Nagasaki was based on plutonium (Pu-239).

2.2 Nuclear Chain Reaction and Nuclear Power

In the fission process, as exemplified above, two or three neutrons are produced. As there are a large number of fission reactions with U-235, the average number of neutrons produced per fission has been estimated to be about 2.4 neutrons. Suppose that two neutrons are produced. Then, those two neutrons will cause the fission of two more U-235 nuclei, if the neutrons collide and are captured successfully. The

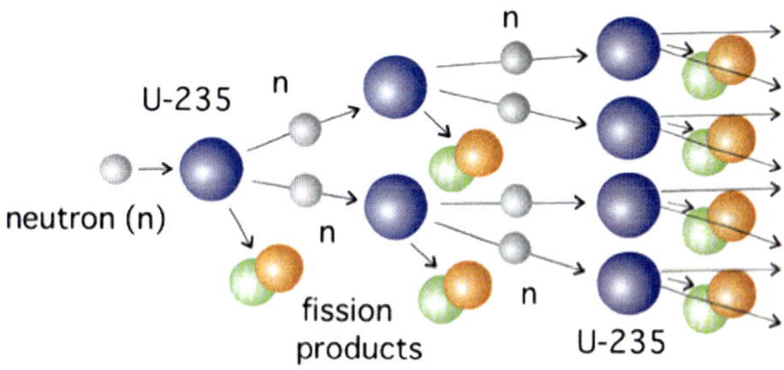

Fig. 2.2 A model of nuclear fission chain reaction (Modified from Fig. 17.24 in Atkins and Jones (2005))

result is the formation of four neutrons, which then fission four more U-235. This process of automatically repeating fission reactions once it starts is called a "chain reaction" (Fig. 2.2).

As mentioned earlier, on average, 2.4 neutrons are produced from one neutron during a fission process. If all 2.4 neutrons hit further U-235 without being lost, then the neutron multiplication factor is $K = 2.4$. If one neutron is lost on average before it hits a U-235, then $K = 1.4$. When $K = 1$, the chain reaction will barely be sustained. When this condition is met, it is said to be at a "critical state". How many neutrons will be lost before colliding with U-235 would depend on the mass of the U-235 sample and other factors. The mass of U-235 at which the critical state is attained is called the "critical mass".

If $K < 1$ (subcritical condition), the fission reaction will soon cease or the fission chain reaction would not commence. On the other hand, if $K > 1$, the fission reaction will be accelerated and soon explode. Clearly, the atomic bomb works under the supercritical condition $K > 1$, whereas in the nuclear reactor, it should be maintained more or less close to the $K = 1$ condition. This is the main difficulty faced in the nuclear reactor.

2.3 Atomic Bombs and Nuclear Reactors

An atomic bomb has to be maintained under subcritical conditions before use. When it is to be exploded, it has to be brought quickly to supercritical conditions. In the atomic bomb dropped on Hiroshima, two separate U-235 masses (under subcritical conditions) were brought together by the action of a gun; one mass was shot towards the other mass. The overall mass then instantly became supercritical and it exploded. The other atomic bomb dropped on Nagasaki was based on plutonium, and the mechanism to bring about a supercritical condition was different (implosion). Atomic bombs require no other consideration technically; all it has to do is bring about a supercritical condition to cause an explosion. No other control was necessary in principle; the only purpose was to explode and destroy everything in its vicinity.

On the other hand, the so-called "peaceful" use of nuclear power requires a strict control of the fission reaction under conditions close to the "critical state". Otherwise, it easily leads to an explosive disaster, as happened in the nuclear facility at Chernobyl in the former Soviet Republic (1986). Several designs of nuclear reactor are based on how to moderate the neutron speed and its number (to control the K-value). This treatise is not intended to discuss the technical details of a nuclear reactor. Therefore, the most commonly used type, i.e., the boiling water reactor, is mentioned briefly here.

Light water (in contrast to heavy (deuterium) water) is used in the core surrounding the nuclear fuel rods (containing enriched U-235), and functions both as the coolant (and energy carrier as vapor) and the neutron moderator. Control rods would be inserted among the fuel rods. The control rod should contain substances that absorb neutrons efficiently. A typical control rod for a light boiling water-type reactor is made of an alloy of silver (Ag), indium (In), and cadmium (Cd). Another is boron (B), particularly $_5B^{10}$, which has a large neutron cross section so that it easily absorbs neutrons; it is commonly used as boron carbide rod. During the process, $_5B^{10}$ turns into another stable isotope, $_5B^{11}$. When the control rods are fully inserted, insufficient neutrons become available, and the fission reaction will cease. So that is how extensively the control rods are to be inserted in order to regulate the fission reaction. This is the most critical device in the nuclear power reactor.

2.4 Radioactive Isotopes and Their Decay

As mentioned earlier, the fission reaction of U-235 leads to the production of a large number (as many as 200) of fission products (daughter nuclei). The mass numbers of these products are divided into two groups; one from about 80 to 110, and the other from 110 to 150. And when two daughter nuclides form, the total mass number of them would be 234–232 or so. In terms of elements (i.e., the number of protons = atomic number in the daughter nuclides), the atomic number of fission products ranges from about 34 (Se) to 62 (Sm). Some of these fission products are listed in Table 2.1 in the order of their yield. The majority of them turned out to be "radioactive". Often, it is usually the case that a nuclear reaction would lead to the formation of nuclei that are not in the most stable form. An unstable, i.e., high energy, state would spontaneously turn into a more stable state. This spontaneous change in a nucleus is called "decay" or "disintegration". The degree of blockage against decay would determine how fast the nucleus changes into a stable state, i.e., its "half-life".

Unstable nuclei may take a number of routes to turn into more stable states. The major paths are: (1) emission of nucleons: protons, neutrons, or α-particles (helium nuclei); (2) emission of electrons (β) or positrons (β^+) (antielectron), and (3) emission of γ-rays (photons).

All the isotopes of the elements higher than atomic number 84 are unstable, and emit α-particles. In these nuclei, there are a sufficiently large number of protons,

Table 2.1 Major thermal nuclear fission products of U-235 and others

Isotope	Yield (%)	Decay mode	Energy (keV)	Half-life[a]	Decay product, comments
Cs-133 → Cs-134	6.79	β	211	2.065 y	Cs-133 is stable, but it captures a neutron to turn into Cs-134, which β-decays, but also captures a neutron to turn into Cs-135
Cs-134		β(γ)	2,059 (605, 796)	2.07 y	→ Ba-134; Ba-134 m1, m2 -γ→
Cs-135		β	269	2.3 My	→ Ba-135
I-135 → Xe-135	6.33	β	135	6.57 h	Neutron capture converts 10–50 % of ^{135}Xe to ^{136}Xe; the remainder decays (9.14 h) to Cs-135 (2.3My)
Zr-93	6.30	β, (γ)	90	1.53 My	→ Nb-93 m -γ (14 years) → Nb-93
Mo-99	6.1	β	1,357	65.94 h	→ Tc-99 m -γ (142 keV, 6 h) → Tc-99
Cs-137	6.09	β, γ	514, 662	30.17 y	→ Ba-137 m -γ → Ba-137
Tc-99	6.05	β	294	211 ky	→ Ru-99
Sr-90	5.75	β	546	28.9 y	→ Y-90
I-131	2.83	β, γ	606, 364	8.02 d	-β → Xe-131 m -γ → Xe-131
Pm-147	2.27	β	224	2.62 y	→ Sm-147
Sm-149	1.09			Stable	
I-129	0.65	β, γ	194, 236	15.7 My	-β → Xe-129 m -γ → Xe-129
Sm-151	0.42	β	77	90 y	Neutron capture to stable Sm-152
Ru-106	0.39	β	39	373.6 d	-β → Rh-106 -β → Pd-106
Kr-85	0.27	β, (γ)	250	10.78 y	→ Rb-85; 0.43 % β/γ (514 keV)
Pd-107	0.16	β	33	6.5 My	→ Ag-107
Se-79	0.051	β	151	327 ky	→ Br-79
Eu-155 → Gd-155	0.033	β	253	4.76 y	Both capture neutrons
Sb-125	0.030	B, (γ)	767 (145)	2.76 y	→ Te-125 m -γ(57.4 d) → Te-125
Sn-126	0.024	β	380	230 ky	→ Sb-126
Gd-157	0.0065			Stable	Capture neutrons
Cd-113	0.0003	β	322	14.1 y	Capture neutrons
K-40[b]		β γ (10 %)	1,311 1,460	1.25 Gy	90 % → Ca-40; 10 % K-40 -EC[c]/γ → Ar-40
U-235[b]		α	4,679	704 My	
U-238[b]		α	4,267	4.51 Gy	
Ra-226[b]		α	4,817	1.6 ky	
Pu-239[d]		α	5,245	24.4 ky	
H-3[e]		β	18.6	12.3 y	

Yield data are from Wikipedia, and others from many different sources
[a]d = days, y = years, ky = 10^3 y, My = 10^6 y, Gy = 10^9 y
[b]These are naturally occurring
[c]Electron capture
[d]This forms from U-238 through neutron capture and two consecutive β-emissions
[e]Is tritium and is not a fission product but forms in a nuclear reactor

which exert destabilizing repulsive effects. This is reflected in the decrease in the nuclear binding energy, as seen in Fig. 1.2. The release of a highly stable α-particle would reduce this tension, and the energy released is then transferred to the kinetic energy of the particle emitted, and this is α-radiation. The kinetic energy of an α-radiation is determined by the energy state of the nucleus and its change, and is independent of the circumstances under which it occurs. Hence, it is possible to identify the nucleus by measuring the radiation energy.

A nucleus that has too many neutrons may reduce its high energy level by either releasing a neutron or converting a neutron to a proton. The former process does not change the atomic number (hence, the identity of the element), and only diminishes the mass number by one. More often, the latter process takes place. When a neutron turns to a proton, an electron is released to compensate for the electric charge (which needs to be conserved), and another particle, an antineutrino, is released alongside. The electron emitted with high kinetic energy is the β-particle. Because of the extra particle, the antineutrino, the kinetic energy carried by the β-particle will vary depending on how the antineutrino behaves. However, the average value (or the distribution) of the energy can be used to determine the identity of the β-emitting isotope.

The β-decay product may not be the most stable state of a nucleus (called the "metastable state" and denoted "m" after the mass number), and, if so, the extra energy would be emitted as a photon, γ-radiation. Hence, β-decay (radiation) is accompanied by γ-radiation in many cases. γ-radiation also occurs along with α-radiation, although not always. As the photon emission depends only on the two energy states of a nucleus, it is characteristic of it. The energy of γ-emission can, thus, be used to identify the isotope.

Some of the representative decay energy values are given in Table 2.1. They range from 20 to 5,200 keV (5.2 MeV). This value should be compared to the chemical energy, the typical value of which is on the order of 1–10 eV. Hence, the nuclear decay energy is several thousand to millions of times as large as the chemical energy.

Table 2.1 includes four naturally occurring radioisotopes, U-235, U-238, Ra-226, and K-40. There are a few other naturally occurring radioisotopes, including Th-230, and the decay products of U-238 and Th-230, such as Ra-226, Rn-222, and Po-220. Pu-239 is an example of transuranium elements that are artificially produced from U or other elements through neutron bombardment, α-particle bombardment, or by other means. Pu-239 occurs via the neutron bombardment of U-238:

$$ {}_{92}U^{238} + {}_{0}n^{1} \rightarrow {}_{92}U^{239} \rightarrow {}_{93}Np^{239}\left(+{}_{-1}e^{0}\right) \rightarrow {}_{94}Pu^{239} + {}_{-1}e^{0} $$

The intensity of radioactivity is measured as how many decay processes take place over a unit time (second). The unit of radioactivity is called the "Becquerel" (Bq). H. Becquerel was the French scientist who first discovered the phenomenon of radiation. Radioactivity used to be expressed in terms of "Curie" (Ci). Marie

Table 2.2 Quantity to represent radioactivity (Bq)

Isotope	Half-life ($t_{1/2}$)	K-value (s^{-1})	No. of atoms to give 1 Bq	Quantity (g) to give 100 Bq
U-235	7×10^8 years	3.14×10^{-17}	3.2×10^{16}	1.2×10^{-3}
U-238	4.5×10^9 years	4.88×10^{-18}	2.1×10^{17}	8.1×10^{-3}
Th-232	1.4×10^{10} years	1.57×10^{-18}	6.4×10^{17}	2.5×10^{-2}
Pu-239	2.4×10^4 years	9.16×10^{-13}	1.1×10^{12}	4.3×10^{-8}
I-131	8 days	1.00×10^{-6}	1.0×10^6	2.2×10^{-14}
Cs-137	30 years	7.33×10^{-10}	1.4×10^9	3.1×10^{-11}
Sr-90	28.8 years	7.36×10^{-10}	1.4×10^9	2.0×10^{-11}
H-3	12.3 years	1.78×10^{-9}	5.6×10^8	9.3×10^{-14}
K-40	1.25×10^9 years	1.76×10^{-17}	5.7×10^{16}	3.8×10^{-4}

Curie further studied the phenomenon and identified the radioactivity caused by radium, and the unit "Ci" was based on a gram of radium. It turned out to be too large a unit, so it was substituted by the Bq, and 1 Ci $= 3.7 \times 10^{10}$ Bq. There is no ambiguity with the activity value Bq, as it is measured as the number of disintegrations per second, as long as it is properly measured.

The decay process obeys the first-order reaction rate equation, so it can be expressed as $-dN/dt = kN$, where k is the rate constant (representing what proportion of the nuclides would decay per second) and N is the number of radioactive nuclei. The $-dN/dt$ value represents "Bq". The integrated form of the rate equation is given by $N = N_0 \exp(-kt)$, where N_0 is the initial quantity. When $t = T$ (half-life), N becomes half of N_0, so that $N = N_0/2$ at $t = T$. Therefore, $\ln 2 = kT$. T (half-life) is used to express how fast the disintegration proceeds. The half-life data for the major radioisotopes found in the fission reaction products are given in Table 2.1.

Radioactivity is usually expressed in terms of Bq/kg of a sample. As the equation above indicates, it represents the quantity N (number of nuclei). For example, in today's Japan (after the Fukushima incident), the allowable radioactivity of food is set at 100 Bq/kg. Such a sample in today's Japan contains Cs-137 as the major component. Let us calculate the quantity of Cs in such a sample of 100 Bq/kg, assuming that it is 100 % Cs-137. The half-life of Cs-137 is 30 years, which is 9.46×10^8 s; therefore, k (rate constant) $= 7.33 \times 10^{-10}\ s^{-1}$. Since 100 Bq = kN, N is 1.36×10^{11}; this many Cs-137 nuclei is 2.26×10^{-13} mol, which is 3.1×10^{-11} g. This is a very small quantity, and cannot be determined by usual chemical analysis. Similar calculations can be done with other radioisotopes of interest, and the results are shown in Table 2.2. The table shows that significant radioactivity is brought about by a very small quantity of radioactive material. It also suggests that the biological effects of radiation should be dealt with in terms of the number of radiation particles and their energies, rather than the ordinary expression of the quantity of material in moles or weight, and of energy in joules, as will be discussed below.

The dose of exposure of a material to radiation is expressed in terms of the energy value J absorbed by the material of a unit quantity (usually 1 kg). One joule deposited in a material is defined as 1 Gray (Gy) (= 1 J/kg). It is a very small quantity in terms of energy in a macroscopic world.

The impact on living organisms is expressed in dose equivalent: Sievert (Sv). It takes into consideration how effective each different radiation would be on a living organism's body, and is expressed as $Sv = Q \times Gy$. It has been officially (ICRP) defined as follows. The Q value for β- and γ-radiation is defined as "1", but $Q = 20$ for α-radiation, as it is more effective in affecting the human body. The dose exposed to neutrons depends on its speed, and Q is thought to be in the range 10–15. These Q values cannot be scientifically accurately determined in terms of expressing the real impact on living organisms, particularly in the so-called "internal exposure". The distinction between "external" and "internal" exposure will be discussed later. The most significant exposure to low-level radiation is "internal", but this was not considered in deciding the Q factors assigned to different radiation modes. The inadequacy of the Sv value in expressing the effects of radiation will be discussed later.

References

Atkins P, Jones L (2005) Chemical principles, the quest for insight, 3rd edn. W. H. Freeman, New York

Bernstein J (2007) Plutonium – a history of the world's most dangerous element. Cornell University Press, Ithaca

Chapter 3
The Chemical World

3.1 The Basics

As the earth formed and was cooled, it became an independent body in the universe. Being "independent" implies that it is materially isolated, and not much input and output of material takes place, though it is open to the universe. The material that comes in includes out-of-the-earth bodies, such as meteorites and some particles, the so-called cosmic rays (which includes α, β, neutron, and proton particles). It is certainly wide open to electromagnetic rays (visible light, ultraviolet to γ-rays) or photons. Some of these particles do reach the surface of the earth, but most of them are screened off (this will be discussed later).

There might have been several thousands of various isotopes on the initial earth. By that time, nuclei had combined with electrons (at low enough temperatures so that they can bind to each other) and formed atoms (and then chemical compounds). Many of the isotopes were not stable, and, therefore, they turned into stable entities as time went by. There are only **264** stable isotopes known in the universe. All the other isotopes are unstable and, hence, radioactive. How fast they change (decay) is governed by their "half-life" (discussed in Chap. 2). As a result, only a very small number of radioactive isotopes have remained up to today; they include U-235, U-238, Th-232 (and their daughter nuclei, such as Ra-226, Rn-222, Po-210), and K-40. All the other material on the earth is made of stable (nonradioactive) isotopes.

The elements up to atomic number 83 (Bi, bismuth) have one or more stable isotopes, but polonium (Po, atomic number 84; a daughter nucleus in the decay process of uranium) and heavier elements do not have stable isotopes, and they are α-emitters (but some emit β and γ particles).

The earth itself, its core, mantle, crust, surface soil, water, and all the living organisms on it are made of stable isotopes. However, some radioactive isotopes are present and even exist in significant amounts on the earth. Potassium K on the earth consists of mostly stable K-39 and K-41, but includes radioactive K-40, which currently comprises about 0.012 % of all potassium. The content of U (U-238 mostly) in the earth's crust is about 1 ppm by weight, and that of Th is about ten

E. Ochiai, *Hiroshima to Fukushima*, Science Policy Reports,
DOI 10.1007/978-3-642-38727-2_3, © Springer-Verlag Berlin Heidelberg 2014

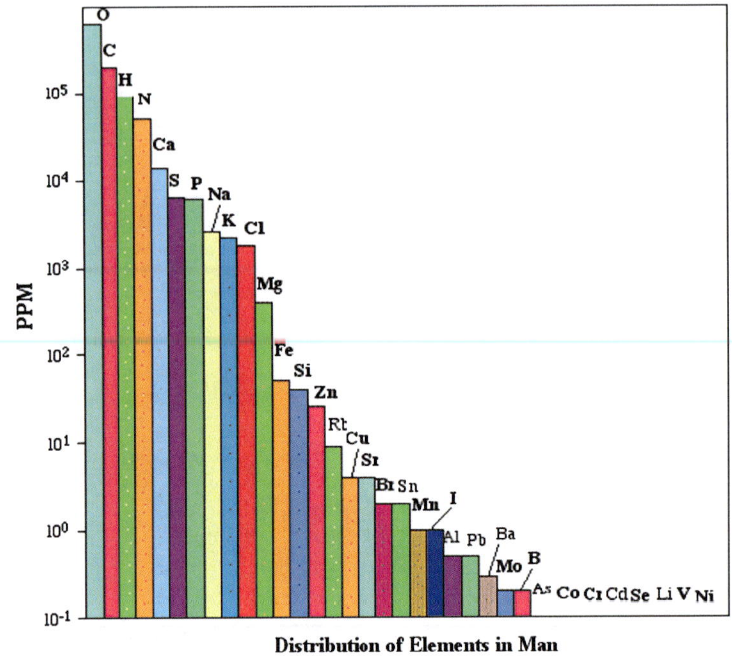

Fig. 3.1 Distribution of elements in man (Fig. 6.1 in Ochiai 2011)

times as much. These elements are present significantly in some kinds of ores and rocks.

Living organisms use about 30 different elements, including H, O, C, N, S, P, Na, K, Ca, Mg, Cl, etc. (see Fig. 3.1). The elements shown in **bold** font are essential for living organisms, and, hence, the organisms have evolved ways to deal with them. They have mechanisms to take up these elements, distribute them throughout the body, incorporate them into specific cells, and then into biocompounds, and eventually eliminate them from the body, and these processes are all chemical reactions (see, for example, Ochiai 2011).

Chemical reactions involve several basic processes: breaking and/or making chemical bonds, and removing and/or adding electrons. In these reactions, the nuclei remain intact, and only the electrons will be moved around: bound together, separated, removed, etc. The energy associated with chemical reactions is based on the electromagnetic forces (electron–electron, electron–nucleus, nucleus–nucleus), and the chemical reaction energies are combinations of energies associated with these processes.

Let us see a few typical examples.

$$CH_4(\text{methane}) : H_3C - H \rightarrow H_3C^{\bullet} + {}^{\bullet}H \qquad (3.1)$$

This is the cleavage of the C–H chemical bond to produce a free radical $H_3C^{\bullet}$ and a hydrogen atom, where the dot "•" represents an electron. The energy to break a chemical bond is known as the "bond energy", and that of the C–H bond is about 412 kJ/mol. This energy is given per mole. One mole of a chemical compound consists of 6.02×10^{23} particles (Avogadro's number). Therefore, the energy required to break a single C–H bond (in methane) is 6.84×10^{-19} J/mol. This is equivalent to 4.27 eV, calculated referring to Sect. 1.5.2.

Ionization refers to reactions such as the following:

$$\begin{aligned} Cu(0) &\rightarrow Cu(+1) + e\,(\text{first ionization}) \\ &\rightarrow Cu(+2) + 2e(\text{second ionization}) \end{aligned} \qquad (3.2)$$

"e" represents an electron. The first ionization energy is 785 kJ/mol, and the second is 1,958 kJ/mol. Again, these are defined for a mole of copper atoms. For a single copper atom, the first and second ionization energies can be calculated to be 8.13 and 20.3 eV, respectively. These are the typical energy changes associated with chemical reactions, and range from *ca.* 1 eV to several tens of eV. This is the magnitude of energy associated with chemical reactions per particle, whether in living organisms or in nonliving material.

It might be interesting to note that visible light (wavelength 320–800 nm) is in the same range as chemical reactions in terms of energy. That is, the middle of the visible range, say, 500 nm, has $\nu = 6 \times 10^{14}\,s^{-1}$ (frequency), and, so, the energy of a photon (500 nm) is 2.48 eV ($E = h\nu$). This is the basic reason for how organisms could have devised mechanisms to recognize visible colors using chemical reactions (physiology). Organisms cannot respond to (i.e., detect) radiation of ranges other than the visible range, though they can detect infrared light as heat. This is one of the basic difficulties in dealing with radiation.

These energy values associated with chemical reactions are to be compared to the energy involved in the nuclear decay processes shown in Table 2.1. The energy associated with radiation is much larger by several orders of magnitude than chemical energies. This point is fundamental to the understanding of the effects of radiation on living organisms, and will be discussed further later.

3.2 Nonliving Chemical World

Everything we see on the earth, the moon, and the other planetary bodies in the solar system, and even some non-visible matter such as air, are made of chemicals. Most of the solid materials are metals, rocks, and organic compounds of high molecular weight. Metals are formed through intermetallic bonds, which is a kind of covalent bond, sharing electrons among atoms. Rocks are usually made of ionic bonds between metallic cations and some anions, e.g., $CaCO_3$ (Ca(+2) and $CO_3(-2)$).

Most organic compounds are made of interatomic essentially covalent bonds, and have relatively weak intermolecular interactions, because such interactions are mostly of the dispersion type. These chemicals are virtually made of stable isotopes, though there is an occasional presence of unstable isotopes, i.e., naturally occurring radioactive isotopes.

Let's look at a sample of 1 kg of a rock, e.g., Mg_2SiO_4, in terms of the number of particles (compound units, molecular units, etc.). 1 kg of it is 7.1 mol of this formula unit, and it consists of 4.3×10^{24} units of Mg_2SiO_4. How about 1 kg of sugar ($C_{12}H_{22}O_{11}$)? Its molar mass is 342 g/mol, and, hence, 1 kg of it contains 2.92 mol and 1.8×10^{24} sugar molecules. 1 kg of lead (Pb) contains 2.9×10^{24} Pb atoms. The significance of these calculations of the number of particles present in a sample will become apparent when the effects of radiation is discussed.

3.3 Living Chemical World

Living systems are essentially a chemical world, though quite a complicated and delicate one. It is complicated in that many thousands of chemicals are interrelated, even in a simple bacterial cell, and that the system in multicellular organisms involves more complex intercellular, intertissue, and interorgan interactions which are physical, physiological, and chemical. It is delicate in that these chemical reactions are intricately balanced so that the overall effects are "living in good, healthy conditions" and "various functions in good order: growth, cell mitosis/ meiosis, etc.". These are in the realm of biochemistry, physiology, and cell biology, and will be dealt with in Chap. 12.

We will briefly look at the magnitudes of the number of molecules involved in a cell of the human body. The cells in the human body vary in size; the average size is estimated to be 4×10^{-9} cm^3, with an average density of 1.15 g/cm^3, and, hence, a single cell weighs about 4.6×10^{-9} g (p. 66, Alberts et al. (2002)). The average cell in the human body is thought to contain 70 % water by weight, proteins 18 %, lipids 5 %, carbohydrates 1 %, nucleic acids (DNA 0.25 % and RNA 1.1 %) 1.35 %, carbohydrate 2 %, small metabolites 3 %, and inorganic ions 1 %. Based on these numbers, we estimate the number of molecules in a cell as follows:

1. 1 kg of a human body consists of 2×10^{11} cells
2. Water in a cell = 3.2×10^{-9} g = 1.1×10^{14} molecules
3. Proteins in a cell = 8.3×10^{-10} g = 4.2×10^{12} molecules of AAs = 8×10^{9} protein molecules (assumption: the average molecular weight of AA = 120 g/mol, and the average protein consists of 500 AAs; if there are 20,000 different proteins, then the average copy of each protein is about 4×10^{5})
4. Nucleic acids in a cell = 2×10^{-13} mol of nucleotides = 1.1×10^{11} nucleotides DNA = 2.2×10^{10} nucleotides; DNA (double strand)

length $= 5 \times 10^9$ nucleotides; RNA $= 9.8 \times 10^{10}$ nucleotides (assumption: the average molecular weight of nucleotides $= 320$ g/mol)
According to the data on p. 202 of Alberts et al. (2002), the length of the DNA in the human genome is 3.2×10^9 nucleotides.

In reality, there are cells of many different sizes and types, and they contain these compounds in various proportions, as well as many other compounds not listed here. But the numbers of molecules in a cell estimated above may give ballpark or typical figures of important component molecules in a cell.

One important cation is often discussed in regard to radiation effects, which is potassium-40, a radioisotope. Potassium K (mostly stable K-39 and K-41) is essential to all organisms and distributes throughout the body, more inside cells than outside. Overall, the content of K in a human body is about 2,200 ppm on average. Therefore, it is about 2.2 g/kg (of body). K-40's natural abundance is 0.0012 % among all potassium. Since a living system operates chemically, it cannot distinguish between K-40 and K-39 (nor K-41), and, hence, it will incorporate K-40 in proportion to the natural abundance. By the way, sodium (Na) is present at about 2,300 ppm and rubidium (Rb) at 9 ppm in the human body. These are three major elements in the alkali metal column in the periodic chart (Fig. 1.3), but Rb is not known to be essential to organisms. This issue will be discussed in detail later (Chap. 10).

References

Alberts B, Johnson A, Lewis J, Raff M, Roberts K, Water P (2002) Molecular biology of the cell, 4th edn. Garland Science, New York

Ochiai E (2011) Chemicals for life and living. Springer Verlag, Heidelberg

Part II
The Military Uses of Nuclear Power: Hiroshima

Nuclear power was initially applied to weapons, as in the tradition of the human race. This part outlines the development of nuclear weapons and how they were used on humankind, and their devastating effects.

Chapter 4
Developing Atomic Bombs: The Manhattan Project

4.1 Birth of the Atomic Bomb

The first application of the nuclear fission reaction was a "weapon of mass destruction" because it produces an enormous amount of energy that could kill a large number of people instantly. Right after the German scientists discovered the nuclear fission reaction in 1938, some scientists associated with the Nazi movement started to develop an atomic bomb. This alerted scientists in the United Kingdom and the United States of America. A. Einstein in cooperation with L. Szilárd, a Hungarian immigrant physicist, wrote a letter (dated August 2, 1939) to President T. Roosevelt that the United States should develop the atomic bomb before Nazi Germany succeeded. Thus, a secret project called the "Manhattan Project" was organized in December 1941 under the leadership of the army general L. Groves and physicist R. Oppenheimer. It involved a large number of scientists, including most of the prominent investigators at the time in the nuclear physics field, and developed most of the basic nuclear industrial processes. It was recognized that either uranium-235 or plutonium-239 can be used to produce such a weapon. A nuclear reactor called Chicago Pile-1was the first one successfully brought to a critical condition ($K = 1.0006$), achieved by E. Fermi and coworkers at the University of Chicago on December 2, 1942. The industrial production of plutonium through the use of a nuclear reactor was carried out at Hanford (started in September, 1943), Washington State, enrichment of uranium-235 at Oak Ridge, Tennessee, and weapons design and others at Lost Alamos, New Mexico. A couple of facilities are shown in Fig. 4.1. It involved major corporations, some of which are still involved in the nuclear industry. The details of the Manhattan Project can be found in Gosling (1999) and http://web.archive.org/web/20101118041329/http://www.cfo.doe.gov/me70/manhattan/index.htm.

Overall, the project spent approximately 2 billion dollars (in 1945 dollar value) and succeeded in producing a Pu-based atomic bomb and tested it successfully on July 16, 1945 (Trinity test at Alamogordo desert, New Mexico). Another completed bomb had already been shipped to the West on July 14, 1945. This was the atomic

E. Ochiai, *Hiroshima to Fukushima*, Science Policy Reports,
DOI 10.1007/978-3-642-38727-2_4,

Fig. 4.1 Hanford B Reactor and K-25 facility at Oak Ridge

bomb dropped on a human population for the first time in human history, on Hiroshima on August 6, 1945 (Fig. 4.2). This bomb was of the enriched U-235 type, and the second bomb was of the Pu type and was dropped on Nagasaki on August 9, 1945 (Fig. 4.3). Japan surrendered unconditionally on August 15 (August 14 in Eastern US time).

It is worthwhile to point out that the 155 scientists involved in the Manhattan Project warned the president that: "The atomic bombs at our disposal represent only the first step in this direction, and there is almost no limit to the destructive power which will become available in the course of their future development. Thus a nation which sets the precedent of using these newly liberated forces of nature for the purposes of destruction may have to bear the responsibility of opening the door to an era of devastation on an unimaginable scale" (cited in Kuznick 2007).

4.2 Why Did the United States Drop Atomic Bombs on Japan?

A controversial issue is why the United States dropped A-bombs on Japan. This issue may not ever be settled. However, two major reasons have been proposed for these actions. One is the issue of accountability of the project. The Manhattan Project was a secret project and cost about 2 billion dollars (in 1945 dollar value). When it was revealed after the war, the government and the project team had to account for the amount of money they spent. They had to show some concrete results for the money being spent well. They needed to drop A-bomb(s) before the end of the war, otherwise, they may have been accused of wasting money—a lot of money. The project director and the military gave a directive to drop a bomb before informing the president, who was in Potsdam, Germany, at that time. They did this despite the fact that the government and the military knew fairly well that Japan would surrender soon without such an atrocious destruction (Global Res 2012).

Another factor is the threat from the Soviet Union. It was threatening to enter the Pacific War, and the United States wanted to prevent it; hence, they tried to hasten the end of the war by dropping an A-bomb. At the same time, the United States wanted to demonstrate the superiority of their military power to the Soviet Union.

Fig. 4.2 Dropping of "Little Boy" on Hiroshima

Fig. 4.3 Dropping of "Fat Man" on Nagasaki

Then why did they drop a second A-bomb on Nagasaki? Wasn't one enough? This question has not been asked very often. A possible reason might be that they wanted to experiment with two different types of A-bomb under real situations, as they produced two bombs at the time, and see the difference. The bomb dropped on Hiroshima was based on U-235 and of a simpler structure, whereas the one dropped on Nagasaki was based on plutonium. The latter type was tested on July 16, 1945, but, perhaps, they wanted to try it under real conditions. Was it "mere scientific curiosity"?

4.3 Nuclear Arms Race

A vigorous nuclear arms race ensued after Hiroshima/Nagasaki, particularly involving the United States and the Soviet Union. The following is a timeline of the chronological development of the nuclear arms race. In addition to the bombs based on nuclear fission, those based on nuclear fusion (that is, what is happening in the Sun) have also been tested. This type is called the "hydrogen bomb", but it has eventually been abandoned.

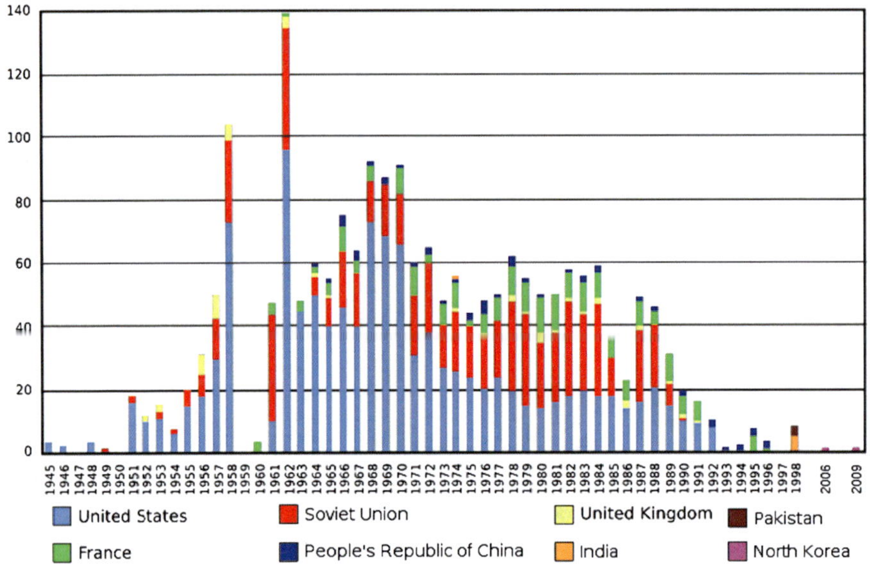

Fig. 4.4 Nuclear explosion tests (from Wikipedia)

1. **US**: first test on 1945.07.16; 1,050 tests so far
2. **USSR** → **Russia**: first test on 1949.08.29; 715 tests
3. **UK**: first test on 1952.10.02; 45 tests
4. **France**: first test on 1960.02.13; 210 tests
5. **China**: first test on 1964.10.16; 44 tests
6. **India**: first test in 1974 (tests in 1998.05 prompted Pakistan); 6 tests
7. **Pakistan**: first test in 1998.05.28; 6 tests
8. **Israel** is believed to have some nuclear arsenals
9. **North Korea**: 3 tests, 2006.10.09, 2009.05.25, 2013.02.12

In addition, a few other countries are said to be in the process of making nuclear arsenals. These nuclear nations have conducted a number of explosion tests above ground and underground; they are summarized in Fig. 4.4. These tests have released an enormous amount of radioactive material around the world. Some of the radioactive isotopes released have very long half-lives and have raised the background level of radiation significantly. Most nuclear nations stopped testing in 1996 when a comprehensive test ban treaty was signed, though it has not been in full force as of 2012. As a matter of fact, the United States conducted a test on November 7, 2012. North Korea has not joined the test ban treaty, and conducted their third nuclear bomb test on February 12, 2013.

References

Global Research (2012) http://www.globalresearch.ca/the-real-reason-america-used-nuclear-weapons-against-japan-it-was-not-to-end-the-war-or-save-lives/5308192

Gosling FG (1999) The Manhattan Project: making the atomic bomb. DOE/MA-0001. History Division, Department of Energy, Washington, DC

Kuznick PJ (2007) The decision to risk the future: Harry Truman, the Atomic Bomb and the Apocalyptic Narrative. http://www.japanfocus.org/-Peter_J_-Kuznick/2479

Chapter 5
Devastation Caused by the Atomic Bombs: Hiroshima and Nagasaki

Now, let's go back to the atomic bombs dropped on Japan. The following is a summary of the features of the atomic bomb ("Little Boy") dropped on Hiroshima. It has been estimated that it was equivalent to 16 KT of TNT (trinitrotoluene, explosive). Only about 860 g of the 70 kg of U-235 contained in the bomb was thought to have exploded. The energy released was 6.3×10^{13} J (63 TJ) (Little Boy, Wikipedia; Los Alamos Report (1985)). About 50 % of it caused the wind-blast, 35 % turned into heat, and 15 % was released as radiation. The temperature just below the epicenter is believed to have reached as high as 6,000 °C. Many charred bodies were scattered around there. The high temperature caused fires. The pressure created by the wind was about 35 atm at the epicenter, and 5 atm even 2 km away from it. This caused a very strong wind (shockwave) that destroyed most buildings and killed many people by the flying debris. Just as a comparison, a hurricane or typhoon's very strong wind is caused by a pressure difference of only about 0.1 atm. These two factors, blast and heat, caused the devastating visible effects of the atomic bomb.

The "Fat Man", dropped on Nagasaki, contained 6.2 kg of highly enriched Pu-239, and about 1.2 kg of it exploded and was converted to energy (Fat Man, Wikipedia). The energy released was estimated to be about 8.3×10^{13} J and equivalent to 22 KT of TNT, 40 % more powerful than the "Little Boy". Figure 5.1 shows the aerial photos of a part of Nagasaki city taken before and after the dropping of the atomic bomb. This enormous change was brought about by a single bomb and almost instantaneously. The area shown in the picture is about 3.5×5 km, and does not represent the entirety of the devastation.

All kinds of radioactive material was released, and, in addition, a very high dose of neutrons reached the land below, making many nonradioactive materials radioactive. In Hiroshima, immediately after the explosion, the so-called "black rain" fell over a large area of 25×19 km, and the amount of rain recorded was 5–100 mm over 3 h. It is thought to have contained radioactive material produced by the explosion, as well as the remaining unexploded uranium. It killed most of the fish in the rivers and ponds, and affected cattle that fed on the contaminated weed, as well as human beings (loss of hair, blood in feces, and other severe symptoms).

E. Ochiai, *Hiroshima to Fukushima*, Science Policy Reports,
DOI 10.1007/978-3-642-38727-2_5, © Springer-Verlag Berlin Heidelberg 2014

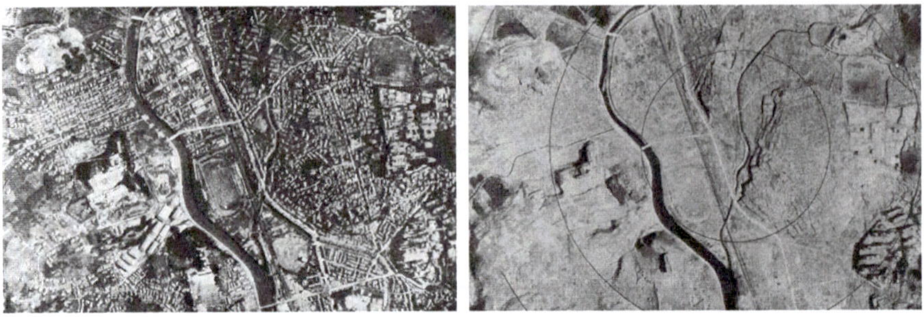

Fig. 5.1 Nagasaki city before the atomic bomb (*left*) and after the bomb (*right*)

Fig. 5.2 A charred body and people with skin hanging from their bodies due to the burning gathering near the epicenter (photos by Y. Ymahata)

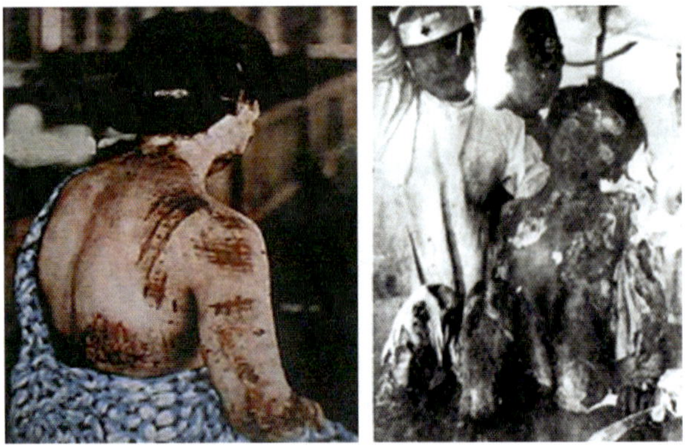

Fig. 5.3 Burned bodies (the photo on the right by Y. Yamahata)

The black rain was recorded in a documentary film "The Effects of Atomic Weapons" produced in 1945 (starting in September) by a Japanese filmmaker, but was classified and never made public. The film also indicated that some soil samples in the area showed high radiation readings (p. 93 in Takahashi (2008)).

The overall number of immediate deaths (by heat and blast) and deaths by radiation in Hiroshima is estimated to have been more than 200,000 by the end of 1945. In comparison, it was about 80,000 in Nagasaki. Some of the damage on human bodies, buildings, and the entire city are seen in Figs. 5.2, 5.3, 5.4, and 5.5.

Fig. 5.4 A cathedral in Nagasaki and the Hiroshima exhibition building destroyed

Fig. 5.5 The instantly flattened Hiroshima

The radiation effects are difficult to visualize, but are also highly devastating and will be discussed later.

References

Los Alamos National report LA-8819. The yields of the Hiroshima and Nagasaki nuclear explosions (Malik J Sept, 1985); http://en.wikipedia.org/wiki/Hiroshima_bomb

Takahashi H (2008) Classified Hiroshima and Nagasaki: US nuclear test and civil defense program (in Japanese). Gaifu-sha, Tokyo

Chapter 6
Other Nuclear Weapons

The atomic bombs mentioned so far are based on the nuclear fission reactions of U-235 or Pu-239. However, other kinds of nuclear weapons have been devised or contemplated. The hydrogen bomb was developed by the United States and the Soviet Union and was tested a number of times, but the projects seem to have been abandoned. It is based on the nuclear fusion reaction (essentially, $4{}_1H^1 \rightarrow {}_2He^4 + 2{}_{+1}e^0$, but more complicated in reality) – that is, the energy source of the sun.

The idea of the neutron bomb has been conceived as a new effective weapon because neutrons penetrate easily into the body, destroy tissues and organs, and also convert most of the stable isotopes in the body into radioactive ones and, hence, kill it. The basic principle of the neutron bomb is the same as for the atomic bomb, so that the blast and heat are unavoidable, but it tries to reduce this portion and increase the proportion of the release of neutrons. Neutrons can penetrate protective materials (armor) or tanks, as well as human bodies. In addition to the United States, the Soviet Union, France, and China have developed weapons based on this principle, but, again, they seem to have abandoned the research eventually (Neutron bomb, Wikipedia).

In the present circumstance, the most serious weapon are the so-called "depleted uranium (DU) munitions". They do not use the nuclear fission reaction, and, hence, do not generate enormous heat; therefore, they are not a nuclear weapon in the strict sense. The uranium from which U-235 has been extracted (in the U-235 enrichment process) is called "depleted uranium". It is not "uranium-depleted". It is still essentially uranium (though mostly U-238), which is an α-emitter. It is quite cheap, as it is a waste product and no longer worthy as a source of fuel. The uranium metal is quite heavy, with a density of 18.95 g/cm^3, which is comparable to gold with a density of 19.28 g/cm^3. Also, it is quite hard, so a bomb composed of it can penetrate the body of a tank and may reach an underground bunker. Hence, such a bomb is very effective and, yet, can be produced cheaply. It has been widely used in Iraq, Yugoslavia, etc. Its health effects are very controversial and will be discussed in Sect. 16.4.

E. Ochiai, *Hiroshima to Fukushima*, Science Policy Reports,
DOI 10.1007/978-3-642-38727-2_6, © Springer-Verlag Berlin Heidelberg 2014

Part III
The “Peaceful” Uses of Nuclear Power: Fukushima

The peaceful use of “atomic” energy has been emphasized and impressed upon the people to reduce the perception of the evilness of the “atom” (bomb). As a result, a large number of nuclear power reactors have been built on this earth. Most of them release radiation, even under normal operational conditions, but serious accidents have inevitably released enormous amounts of radioactive material over the planet. Several major incidents will be reviewed here.

Chapter 7
The Development of Nuclear Power Reactors and the Nuclear Industry

7.1 General Development

There was one more military step prior to the "peaceful" application of nuclear power. The United States Navy was not involved in the Manhattan Project, but wanted to use the idea of nuclear energy to power a submarine, so that it could operate for a long period under water. Admiral H. Rickover, the chief engineer of the USS Nautilus' nuclear power plant, adopted the Westinghouse proposal of a "pressurized water reactor" (PWR) and successfully built "S1W" in the 1950s in Idaho. USS Nautilus, the first nuclear submarine, was launched in 1954 (USS Nautilus, Wikipedia). Rickover was then involved in developing commercial land-based nuclear reactors.

The first practical nuclear reactor was designed for use in submarines, where space is quite limited, and security is of utmost importance. All submarines adopted the use of a PWR, a typical example of which is the reactor TMI-2, which is shown in Fig. 8.2. GE developed a simpler design: a "boiling water reactor" (BWR). Many original designers were aware of the various design problems of reactors of this type, but were forced to resign when they made their observations known. The cooling system was particularly vulnerable. Yet, this same design, as well as the PWR type mentioned earlier, was used for the majority of commercial nuclear power plants. The first commercial nuclear plant (of the PWR type) was approved by the Atomic Energy Commission (AEC) in 1953 in the United States, built at Shippingport, Pennsylvania, and went into operation at the end of 1957. The nuclear reactor of the Fukushima power plant is of the BWR type (Hall 2011; Fukushima Dai-ichi nuclear disaster, Wikipedia). That is, the "nuclear power generator can be said to be a direct descendant of a military device".

There is another military connection. Many countries (and both their industry and government), including even Japan, were interested in "peaceful" nuclear power, with an eye on the possible military applications when the need arises. The ordinary nuclear reactor based on U-235 produces a lot of Pu-239 (240, etc.), which can be used to produce atomic bombs. There were also economic reasons for

E. Ochiai, *Hiroshima to Fukushima*, Science Policy Reports,

DOI 10.1007/978-3-642-38727-2_7,

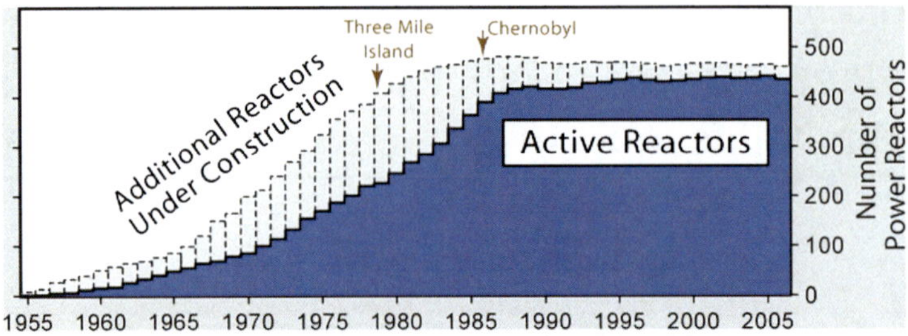

Fig. 7.1 Change in the number of nuclear reactors (Nuclear power, Wikipedia)

promoting nuclear power generation. That is, corporations involved in nuclear weapon developments wanted to use and extend the technologies and know-how obtained in the process for the purpose of profit. They cooperated with Admiral Rickover in developing commercial nuclear power generators. The military interest is, thus, intimately intertwined with the economic interest in the nuclear industry.

In recent years, nuclear power has been touted as a green energy source. In other words, it is reputed that it would not produce greenhouse gas carbon dioxide, such that it is the best solution for combating the problem of global warming. This is entirely false. It is true that it would not produce CO_2 in the process of making electricity. However, if the whole operation of nuclear power generation is taken into account, it produces as much CO_2 as energy production based on fossil fuels. Besides, it releases more than half of the heat produced into the environment as waste, which directly heats up the earth (refer to the note at the end of Sect. 8.1).

Figure 7.1 shows how many nuclear power plants have been built the last 60 years. A list of all nuclear power facilities can be found at http://en.wikipedia.org/wiki/List_of_nuclear_reactors. Since 1990, the world has become cautious in building nuclear power plants, perhaps because of the disaster at the nuclear power plant in Chernobyl in the former USSR. Currently, there are about 440 nuclear power reactors on this earth. They are spread mostly across the northern hemisphere, and many of them are located in earthquake zones (Fig. 7.2). And now, because of the Fukushima accident (2011.03.11), many countries have become cautious in building and operating nuclear power plants. Indeed, as of August 9, 2012, the US government decided to stop issuing permits for new nuclear power plants and license extensions for existing facilities until it resolves issues concerning the storage of radioactive waste. This decision was partly due to the failure in developing long-term storage facilities in Yucca Mountain, Utah.

There are several different types of nuclear reactor. The most widely used is based on light water as the coolant and neutron moderator. Two types of reactor use light water: BWR (Fig. 7.3) and PWR (an example is the Three Mile Island #2 reactor, Fig. 8.2). These require enriched uranium (up to 3 % in U-235) as the fuel. Heavy water (D_2O) is used as the coolant and moderator in "heavy water type" reactors, in which uranium can be used without enrichment, because heavy water is weak in moderating neutrons. By the way, natural uranium consists of 0.0055 % U-234, 0.72 % U-235, and 99.275 % U-238. There are other types of reactors, but they are omitted here.

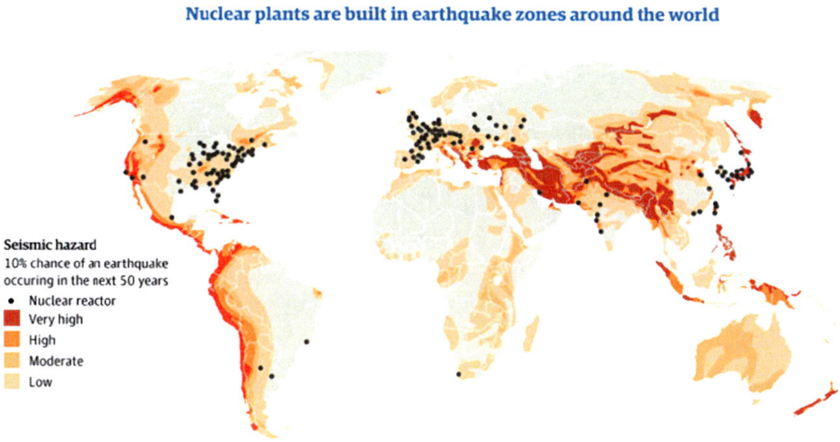

Fig. 7.2 Distribution of nuclear power plants on the earth (http://www.guardian.co.uk/news/datablog/2011/mar/18/nuclear-reactors-power-stations-world-list-map)

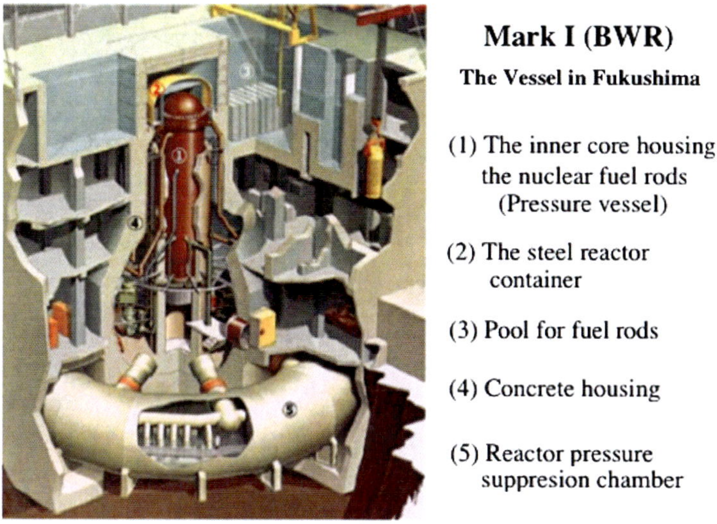

Fig. 7.3 Nuclear reactor of the BWR type

7.2 Nuclear Power Industry in Japan

The Japanese people in the post World War II period were quite averse to the idea of anything "atomic or nuclear", because of their experience of atomic bombs in Hiroshima and Nagasaki. Often, their sensitivity was termed as "nuclear allergy".

On December 8, 1953, President Eisenhower delivered a speech entitled "Atoms for Peace" at the United Nations General Assembly. Its purpose was to reduce the evil image of "atom" or "nuclear" (bomb), and promoted the idea of the "peaceful use" and co-existence of humankind with the nuclear industry. However, right after the speech, in March 1954, there was another nuclear incident with Daigo (#5) Fukuryumaru (literally, "lucky dragon"), a Japanese fishing boat. The fishermen on that boat were exposed to the fallout of a hydrogen bomb test at Bikini Atoll in the South Pacific. One of the fishermen later died. This enraged the Japanese.

Then, the United States government and nuclear industry started an all-out effort to appease the Japanese people through the US Information Service (USIS). They spread the words and virtues of the peaceful use of nuclear energy and the safety of nuclear power plants. The majority of the Japanese people had been persuaded that nuclear power was clean and safe. Unfortunately, many of the atomic bomb survivors did tolerate the idea of energy created by nuclear power as a symbol of "peaceful use".

The US government and industry had offered assistance in building nuclear reactors, transferring the techniques involved, and also providing the enriched uranium for the fuel. They persuaded Mr. S. Matsukata, the then CEO of the newspaper "Yomiuri", and a cabinet member, Y. Nakasone. Nakasone managed to put the first budget for building a nuclear reactor through the legislature in March 1954 just before the #5 Fukuryumaru incident mentioned earlier.

The following brief history of the nuclear power industry in Japan is based on http://www.rist.or.jp/atomica/data/dat_detail.php?Title_Key=16-03-04-01. The first reactor, coined the Japan Power Demonstration Reactor (JPDR), was built in the village of Tokai. It was experimental, of BWR type, and was introduced from the US (GE). It went into operation on October 26, 1963. Its operation was marred with a number of defects and deficiencies. It provided an opportunity to improve the reactor design, so it was claimed. A second experimental reactor of a larger size was abandoned due to having numerous defects. It was decommissioned in March 1976 and dismantled. The dismantling was completed in March 1996, taking a total of 20 years.

Another plant, Tokai Electric Power, was introduced from the UK and started commercial operation in July 1966. The unit1 reactor (BWR type) of the Fukushima Dai-ichi nuclear power plant (of TEPCo) went into operation in March 1971. The accident on 2011.03.11 took place virtually on the 40th anniversary of the commencement of its operation. The oil embargo by the OPEC countries in 1973 facilitated the introduction of more nuclear power plants. By the time of the accident at the Fukushima plant, 54 reactors (mostly BWR or PWR types) were operating in Japan.

On 2012.05.05, all Japanese nuclear reactors had been shut down for one reason or another. However, two reactors in the Ohi plant (in Fukui prefecture, north of Kyoto) were brought back online, despite a very strong opposition from the Japanese people and also in spite of the fact that nuclear power was not necessary in order to maintain an adequate supply of electricity. This was decided politically by the government, but the pressure to do so came from the Kansai Electric Power

Company, which was afraid of going under if their nuclear facilities became toxic assets. This is a major concern of the owners of nuclear facilities, but it will not be dealt with in this treatise.

The other pressure to continue to operate nuclear reactors in Japan is to keep producing plutonium, which can be utilized to make nuclear weapons. This seems to be a real motivation in some segments of the Japanese government pressured by the US authorities.

7.3 General Problems

The major nuclear industrial countries include the US, France, UK, Russia, Japan, and Korea, and other countries such as China and India are trying to increase their reliance on nuclear energy. It is not a purpose of this treatment to pursue this matter any further. However, it must be pointed out that the major issue in this treatment, i.e., radiation effects, should be a concern everywhere around the world. No country can escape the ill effects of radioactive material produced in the process of nuclear power generation.

Radioactive material produced would disappear only in their own time, i.e., according to their half-lives. The waste products from nuclear power reactors are full of radioactive substances, some of which have very long half-lives. An estimate of how long it would take for the radioactivity of the waste to decrease is shown in Fig. 7.4. The overall radioactivity takes at least 2×10^3 years (20 centuries) to diminish to a level equivalent to the natural uranium level. Transuranium nuclides last longer. The initial amount of radioactivity from one ton of spent fuel is estimated to be about 2×10^4 TBq ($T = 10^{12}$). It is a very difficult issue to find ways to dispose of and store these radioactive wastes safely for this long on this planet. If mankind had not invented nuclear power, radioactivity would have remained at the level shown in the middle of Fig. 7.4, and nobody would have to worry about any significant health effects.

Studies are being conducted to convert radionuclides of long life into nuclides of shorter life as a way to make their long-term storage easier. One such method that is being developed is the accelerator-driven system (ADS), in which neutrons created by bombarding protons on a metal target (e.g., Pb-Bi) by an accelerator will fission transuranium nuclides of long lives to daughter nuclides of short half-lives. This method is dependent on nuclear fission reactions, but the process is conducted under subcritical conditions, so that it can be halted by stopping the accelerator (http://nsed.jaea.go.jp/ndre/ndre3/trans/src-e/main.html#ads).

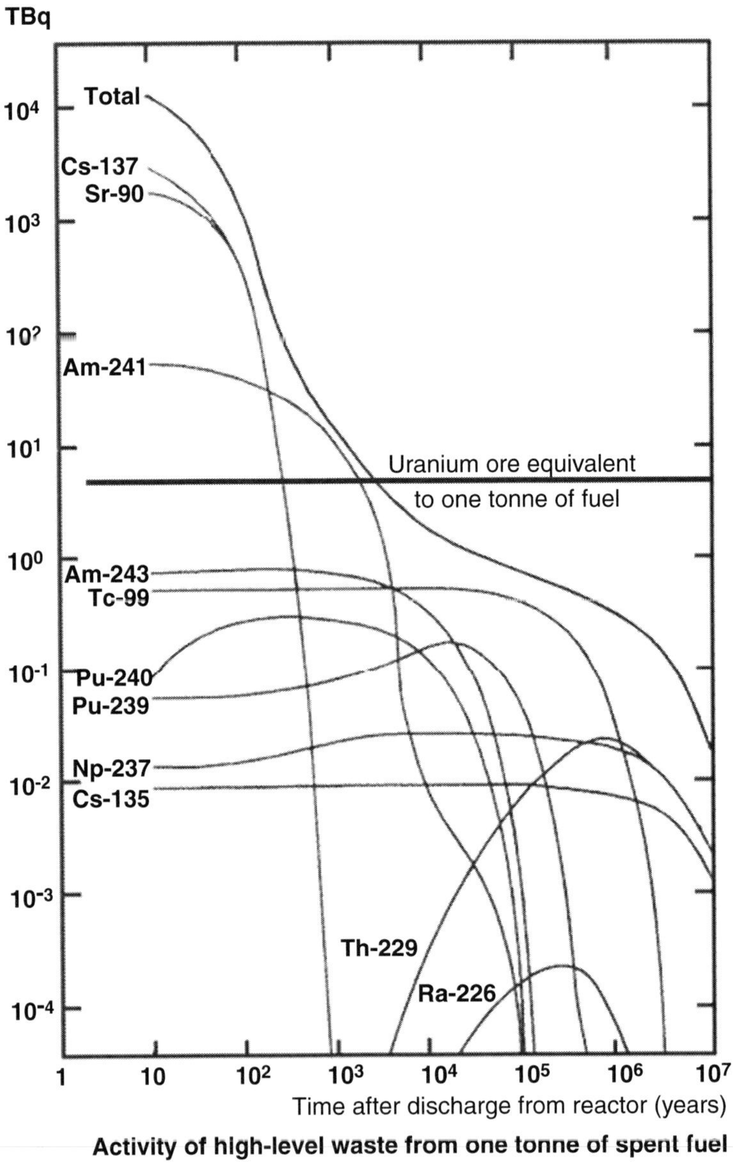

Fig. 7.4 How long the radioactive wastes last

Reference

Hall AJ (2011) "From Hiroshima to Fukushima", http://www.veteranstoday.com/2011/03/28/from-hiroshima-to-fukushima-1945-2011

Chapter 8
Major Incidences of the Failure of Nuclear Facilities

8.1 Radiation Dispersal Under Normal Conditions

Nuclear power plants are vulnerable to accidental failures, due to the basic difficulty in controlling the nuclear fission reaction and regulating the temperature of the fuel rods, and the complicated and multipart structure. An accident, if sufficiently serious, inevitably leads to the release of a large amount of radioactive material. However, nuclear power plants and their associated operations (uranium mining, enrichment, recycling, etc.) release radioactive materials even under the normal, routine operation.

Let us first examine the radiation problems associated with the normal operation of a nuclear power plant. Uranium has to be mined, and it is mostly U-238 (plus some U-235 and U-234), which is an α-emitter of a very long half-life (*ca.* 4.5 billion years). It decays first to thorium-234, then some others, then radium-226, radon-222, etc. All of these are radioactive, emitting α- or β-particles and γ-rays, and are contained in the uranium ores, though radon is a gas and, hence, floats in the atmosphere. Uranium mineworkers are exposed to these radioactive materials, unless they are very well protected against them. Many of the workers in earlier times died of lung cancer, etc., and the situation seems not to have been improved much. The uranium obtained needs to be processed and then subject to the enrichment process. It then has to be molded into fuel pellets and then into a rod. After use in a reactor, the fuel rods will be reprocessed. The workers are potentially exposed to radiation from the uranium throughout the whole process, unless strictly safeguarded. The part of uranium ore discarded is often piled up as tailings (Fig. 8.1; see, e.g., http://www.energy-net.org/01NUKE/u-mining.htm), which are radioactive and can affect people passing nearby, and be washed by rain into the surrounding soil and then underground aquifers or blown away if dried (Caldicott 2011).

A 1,000 MW (=1 million KW) plant (of nuclear power) is said to create, usually, an amount of radioactive material equivalent to that released by the explosion of 1,000 Hiroshima bombs per year (see the NOTE at the end of this section). Under

E. Ochiai, *Hiroshima to Fukushima*, Science Policy Reports,
DOI 10.1007/978-3-642-38727-2_8, © Springer-Verlag Berlin Heidelberg 2014

Fig. 8.1 Tailings of uranium mining/milling operations (Southwestern USA)

normal operating conditions, the reactor needs to be vented occasionally in order to control the pressure inside; the gas released contains radioactive material, though a device to remove radioactive material is supposed to be attached.

The nuclear fuel needs to be reprocessed to extract the remaining U-235 and useful products, such as Pu-239. This requires a very corrosive acid, and is a very dirty process. Workers at every step in the nuclear industry are in danger of exposure to radiation. The whole issue of how and where to store the radioactive (over millions to billions of years) waste material from spent fuel rods has not been solved satisfactorily. Even several decades to several centuries are very long in the normal sense of mankind.

How about other uses of radioactive material, such as medicine, research, etc.? They certainly produce radioactive waste, though usually of a low quantity. The waste disposal of these low-level radioactive wastes, including those from nuclear power plants, was "dumping of the nuclear waste enclosed in metal/concrete drum cans into the deep ocean" in earlier days. The purpose was to eventually disperse and dilute the radioactivity, rather than contain it. This had been practiced by most of the countries that operated nuclear power plants, until it was banned by an international treaty in 1993, though it seems to have been stopped by 1982 (Ocean disposal of radioactive waste, Wikipedia).

This short account on nuclear power production implies that radiation is a serious problem, with hazardous radiation even under normal operational conditions. Subtle effects likely due to the escape of radioactive material from nuclear reactors under normal operating conditions have been observed and reported. These will be discussed later.

[**NOTE:** A nuclear power plant producing 1 million KW (or 1,000 MW $= 10^9$ W) burns about 3–4 kg of U-235 a day. 10^9 W is 10^9 J/s, which amounts to 8.6×10^{13} J/day. 3.5 kg/day of U-235 will produce about 2.3×10^{14} J/day. The "Little Boy" produced about 6.3×10^{13} J. Therefore, a 1,000 MW nuclear power plant produces amounts of energy and radioactive material equivalent to 3–4 Hiroshima A-bombs/day or 1,000 Hiroshima A-bombs per year. These data also indicate that only 8.6×10^{13} J of the total heat energy of 2.3×10^{14} J is converted into electrical energy; that is, the efficiency is about a third. The rest of the heat, i.e., *ca.* two-thirds, is released into the environment as waste. This contributes directly to heating the planet.]

8.2 A Brief History of Major Incidents of Radiation Release

A short history as a list of the major incidents of radiation release at nuclear power plants and related facilities is given below. These are cited mostly from Chugoku Newspaper (1992).

- 1940s on: A high incidence of lung cancer among U-mining workers; it continues even today.
- 1940–1950s: Hanford (Washington) facility released 2×10^{16} Bq of I-131, including an intentional release of radioactive material of spent fuel on 1949.12.02.
- 1954: US H-bomb test; the crews of #5 Fukuryumaru were exposed to the radioactive fallout; one died.
- 1957: An explosion at a reprocessing facility (at Kyshtym) in Ural, USSR; believed to be at level 6 of severity (one notch below the highest). See "ICRP publication 111" for some details.
- 1958: Accident at an experimental reactor in Yugoslavia (one death); at Los Alamos (five exposed and one death).
- 1961: An explosion of a reactor in Idaho (three deaths).
- 1965: A high incidence of abnormal thyroid conditions was found among young people in Utah, downwind from the Nevada test site.
- 1966: Dropping of bombs on Spain: two US military planes collided above Palomares in southeast Spain. Four thermonuclear weapons and men fell around the town. Two of the weapons landed safely, but the parachutes of the other two failed to deploy, and the high explosives in them detonated and burned, though no nuclear explosion took place. Pu-239, 240 contaminated the surrounding area (http://en.wikipedia.org/wiki/1966_Palomares_B-52_crash).
- 1966: USSR nuclear submarine leaked radioactive material (one death).
- 1968: 90 % of children on the Marshall Islands were found to have thyroid abnormality (due to the H-bomb test in 1954).
- 1970: 12.18: Huge mushroom cloud near the Nevada test site (many were exposed and there were two death).
- 1971: The three pilots who flew over the Nevada test site in the 1950s died: two with leukemia and one with cancer; abnormally high incidences of leukemia in Utah and Arizona.
- 1973: 100,000 gal of radioactive waste liquid leaked from the Hanford facility (leakage amounted to 420 million gal over the years).
- 1973: Nuclear fission products leaked at the Sellafield (Windscale) nuclear plant in the UK (since the 1950s).

- 1977: High incidence of leukemia was noted among 170,000 US army personnel who participated in nuclear tests in Nevada.
- 1979.03.28: **Accident at the Three Mile Island nuclear power plant** (US, **Level 6**)
- 1986.04.26: **Chernobyl accident** (USSR, **level 7)**
- 2011.03.11: **Fukushima accident** (Japan, **level 7)**

These are only the major incidents. There have been innumerable accidents, reported or unreported. The major incidents of the failure of nuclear power plants, i.e., the last three accidents above, will be briefly described below.

8.3 The Accident at Three Mile Island Nuclear Power Plant

An almost brand new #2 reactor of about 900 MW capacity at the Three Mile Island nuclear facility on Susquehanna River, Pennsylvania, USA, had an accident in 1979. The accident's outline follows, based on http://www.world-nuclear.org/info/inf36.html.

The reactor was of the PWR type, and is sketched in Fig. 8.2. At 4 a.m. on March 28, 1979, it was operating at full capacity. A minor malfunction in the secondary cooling circuit caused an increase of temperature in the primary coolant. That triggered the system to shut down automatically. A relief valve did open as it was supposed to, but failed to close down, and, as a result, much of the primary coolant drained away. Unfortunately, an instrument failed to indicate this behavior. The result was a partial meltdown of the core. The operators could not control the high pressure and high temperature of the cooling system through March 30. Some radioactive gases leaked from the cooling system. They were believed to be mostly noble gases, and the amount released was estimated to be about 480 PBq (P = peta $= 10^{15}$) in total, including 480–630 GBq of I-131. This was based on inadequate measurements (Caldicott 2011). Now, it is believed that these official figures were grossly underestimated, and a more recent estimate indicated that at least 2.4 PBq of I-131 was also released (quoted in Caldicott 2011). Besides, there was no attempt to measure other α- and β-emitting radioisotopes. In addition, 5,000 l of highly contaminated wastewater was released into Susquehanna River.

Officially, it has been reported that no serious health effects due to the released radioactive material have been identified. After a month, on April 27, the reactor was restored to a normal (below 100 °C) temperature. It turned out that about half of the core had melted. Fortunately, the damage caused by the melted fuel was not too serious. Yet, it took about 12 years to clean up the damaged reactor, which cost about a billion dollars.

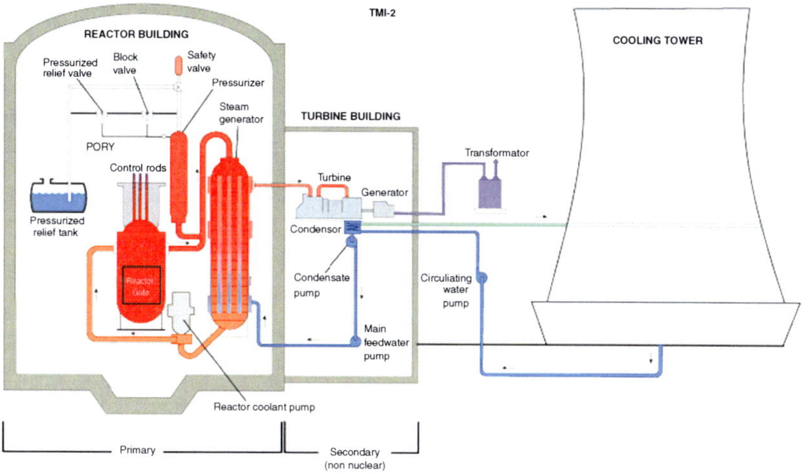

Fig. 8.2 The #2 reactor (PWR) at the Three Mile Island power plant (http://www.nrc.gov/reading-rm/doc-collections/fact-sheets/3mile-isle.html)

8.4 The Accident at Chernobyl Nuclear Power Plant

The most serious accident in human history occurred at the Chernobyl nuclear power plant in 1986 in the Soviet Union (today's Ukraine). It is located near the border between Ukraine and Belarus, and about 130 km north of Kiev, the capital of Ukraine. The reactor was of a type called "RBMK" (reaktor bolshoi moshchnosty kanalny), which has channels of cooling water and uses graphite (solid) as a moderator. The accident of the #4 reactor of Chernobyl will be sketched based on http://www.world-nuclear.org/info/chernobyl/inf07.html.

A test was conducted to see how long the turbine spins and supplies power to the main circulating pumps upon a loss of main electrical power, starting early in the morning of April 26. To start the test, the operator disabled the automatic shutdown mechanism. The operator then tried to manually shut down by inserting the control rods. But somehow, the power of the reactor instead surged, and its interaction with water led to a rapid increase of pressure, causing substantial damage to the fuel and the reactor. Eventually, it led to a severe steam explosion, and, likely, a small-scale nuclear explosion followed, which released fission products into the atmosphere. The accident was made worse by the graphite being on fire. It released further radioactive material. The damaged reactor #4 is shown in Fig. 8.3.

Many people died as a result of these explosions and fires, although the exact number is not known. It has been estimated that a total of about 14×10^{18} Bq of radioactivity was released; over half of it is believed to have been radioactive noble gases (krypton and xenon). Two to five days later, 5,000 t of boron, clay, dolomite,

Fig. 8.3 Chernobyl #4 reactor accident

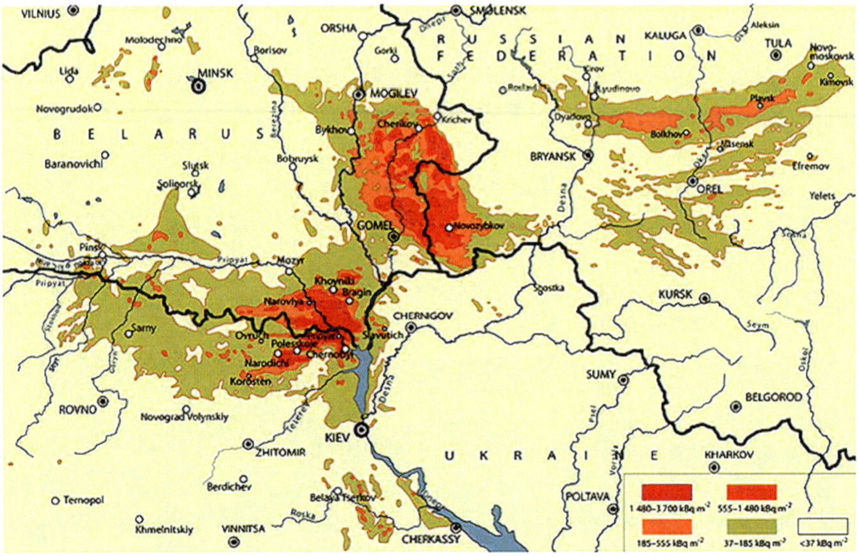

Fig. 8.4 Contaminated areas due to the Chernobyl accident in Ukraine, Belarus, Russia (International Advisory Committee. The International Chernobyl Project. Technical Report. IAEA, Vienna (1991))

lead, and sand were dropped onto the burning core from helicopters, in an effort to extinguish the blaze and limit the radioactivity release.

The radiation dose imposed on the firefighters who worked there on the first day is estimated to have been about 20 Sv, causing 28 deaths by the end of July 1986. The radioactive material from the reactor spread far and wide, particularly toward the northwest (Scandinavia) and west toward Germany and others (see Figs. 8.4 and 8.5).

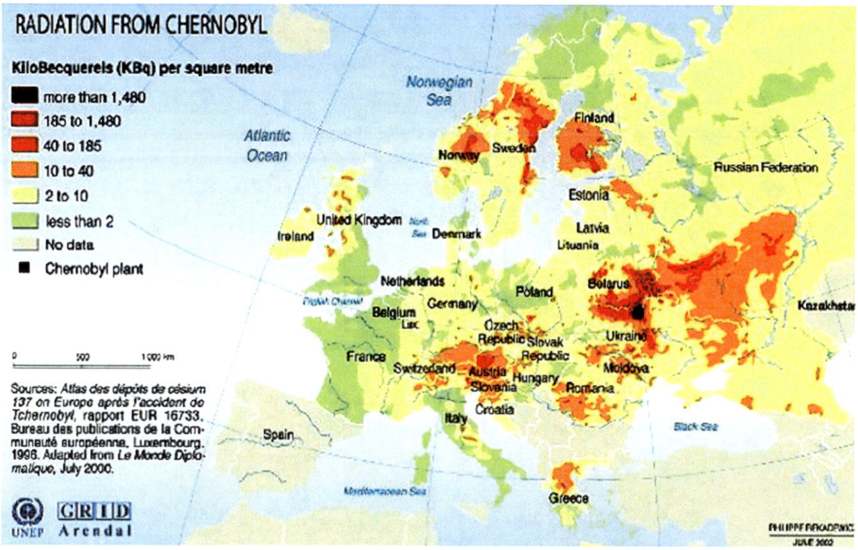

Fig. 8.5 Radiation contamination in Europe due to the Chernobyl accident (Bureau de publications de la Communaute europeenee, Luxembourg)

About 200,000 people were recruited from all over the Soviet Union to clean up the mess in 1986–1987. These people were called "liquidators" and were exposed to high doses of radiation, averaging about 100 mSv; about 10 % of them received *ca.* 250 mSv and a few were exposed to more than 500 mSv. Later, the number of liquidators was increased to 600,000. The initial radiation exposure was mostly I-131, but later, it was Cs-137.

The amount released has been estimated to be 1.8×10^{18} Bq (1.8 EBq) of I-131 (E = exa = 10^{18}) and 8.5×10^{15} Bq (8.5 PBq) of Cs-137. The health effects of the radiation will be discussed later. About five million people lived in the areas contaminated (above 37 kBq/m^2 of Cs-137), and 400,000 lived in more strictly controlled areas (above 555 kBq/m^2).

Some of the visible effects of the Chernobyl incident are shown in Figs. 8.6 and 8.7. Many villages and towns became uninhabitable due to the contamination by radioactive material and were deserted or obliterated (Fig. 8.6). It is said that it takes a couple of thousand years for the immediate area to become inhabitable again. One visible health effect is disfigured children born from the exposed mothers (Fig. 8.7).

Fig. 8.6 Many villages were forced to disappear (to decontaminate)

Fig. 8.7 Malformed babies were born

8.5 The Accident at Fukushima Dai-ichi Nuclear Power Plant

8.5.1 The Causes and Processes of the Accident

The accidents in the last two sections were caused by malfunctions of the reactor system and mistakes made by the operators, and not by any external causes. The Western media blamed the design faults in the RBMK reactor as a cause of the accident at Chernobyl. They said that the Soviet Union, isolated from the West, did not have enough expertise to build better reactors, and, hence, a disaster such as that of Chernobyl would not have happened with the reactors of the Western countries.

However, it did happen with one of the most widely used designs, BWR, in the West, though the reactors were located in Japan; it is the Fukushima incident.

A very strong earthquake (of magnitude 9) shook the northeastern part of Japan's main island on 2011.03.11. Soon after, a huge tsunami as high as 30 m hit the eastern coast of the same area. It devastated the villages, towns, and cities along the coast, and many lives were lost.

There were six reactors in the Fukushima Dai-ichi (No. 1) Nuclear Power Plant (NPP) of Tokyo Electric Power Co (TEPCo), located about 250 km northeast of Tokyo, the capital of Japan. A detailed account of this NPP and the accidents is given on Wikipedia (Fukushima, wiki). A detailed report of the accident investigation by TEPCo itself was published in July 2012, and can be found at http://www.tepco.co.jp/en/press/corp-com/release/2012/1205638_1870.html. Just south of this plant, there is another NPP, Fukushima Dai-ni (No. 2) (TEPCo). No significant accident happened at the Dai-ni plant. Out of the six reactors, units 1–3 in the Dai-ichi plant were operating at the time of the earthquake; unit 4 is said to have been in the waiting stage, storing a large number of spent and new fuel rods (totaling about 1,600) in its storage pool. Units 5 and 6 were not operating. It is believed that the control rods were inserted automatically at the instance of the quake and successfully stopped the nuclear fission reactions in units 1–3. Accidents occurred in all four reactors of units 1–4. The damage to the reactors of units 1–4 was severe and the radiation around the reactors is still so strong that no detailed investigation by humans is yet possible.

The following is, therefore, an incomplete picture of what happened; a reasonable analysis based on many reports. TEPCo insisted that the accidents were caused by the huge tsunami that was much larger and more severe than planned for. However, it seems that the initial problems which led to the disaster were caused by the earthquake. This point is still in dispute. The strong quake displaced, dislodged, or broke several pipes and valves, particularly those which were part of the cooling system. Apparently, there was a significant sinking of the ground beneath the reactors (this was admitted publicly by TEPCo on Aug. 15 th, 2013). The quake caused the failure of the main electrical source. In addition to that, the tsunami also disabled the supplementary emergency power supplies. No power was left to operate the cooling system. One serious mistake was that the emergency cooling system which was operable without needing electricity was not switched on, and no operators or officials noticed that fact. The lack of sufficient cooling water caused overheating of the fuel rods, which led to their melting, the so-called meltdown. This happened first at nuclear reactor unit 1, but it seems to have happened to both reactors at units 2 and 3, as TEPCo later admitted. In the case of the unit 1 reactor, the molten metal, particularly the zirconium metal alloy that wraps the uranium oxide pellets, reacted with water, decomposing it to produce hydrogen gas at high temperatures. Chemically, it is written as $Zr + 2H_2O \rightarrow ZrO_2 + 2H_2$. Hydrogen gas can be ignited and explode when it comes into contact with oxygen. This is the so-called "hydrogen explosion" and is what happened in the unit 1 reactor (http://www.youtube.com/watch?v=VLJspb_mo-s). The radiation level inside the reactor vessel of unit 1 was measured at 11 Sv/h on 2012.10.10, 19 months after the explosion; this level would kill a person in an hour.

Table 8.1 Estimates of the radioactive material released from the reactors of units 1–3 during 3.11 ~ 3.31, 2011 (Data from #4 reactor are not reported here)

Nuclide	Unit 1 reactor	Unit 2 reactor	Unit 3 reactor	Total
Xe-133	3.4×10^{18}	3.5×10^{18}	4.3×10^{18}	1.1×10^{19}
I-131	1.2×10^{16}	1.4×10^{17}	7.0×10^{15}	1.6×10^{17}[a]
Cs-134/137	1.3×10^{15}	3.0×10^{16}	1.5×10^{15}	3.3×10^{16}
Te-127 m/129 m	9.7×10^{14}	3.2×10^{15}	2.8×10^{14}	4.4×10^{15}
Ba-140	1.3×10^{14}	1.1×10^{15}	1.9×10^{15}	3.2×10^{15}
Sr-89/90	8.8×10^{13}	7.3×10^{15}	1.3×10^{15}	2.1×10^{15}
I-132/133/135	1.7×10^{15}	3.7×10^{12}	6.8×10^{13}	1.8×10^{15}
Sb-129	1.6×10^{14}	8.9×10^{10}	3.0×10^{12}	1.6×10^{14}
Np-239	3.7×10^{12}	7.1×10^{13}	1.4×10^{12}	7.6×10^{13}
Ce-141/144	7.7×10^{11}	2.8×10^{13}	3.6×10^{11}	2.9×10^{13}
Zr-95	4.6×10^{11}	1.6×10^{13}	2.2×10^{11}	1.7×10^{13}
Pu-241	3.5×10^{10}	1.2×10^{12}	1.6×10^{10}	1.2×10^{12}
Pu-239/240	1.7×10^{07}	6.1×10^{09}	8.0×10^{07}	6.4×10^{09}

[a]A more recent estimate is 5×10^{17} Bq

The unit 2 reactor is believed to have also undergone meltdown, but no strong explosion was observed. It boiled water vigorously, which escaped from the damaged ring adapter at the head of the reactor. Attempts to vent it to relieve the high pressure were unsuccessful. The SR valve was intended to relieve pressure by transferring water to the pressure control vessel at the bottom. This facilitated meltdown of the fuel rods. The radiation from the unit 2 reactor seems to have been the highest among those from the reactors of units 1–4 (Table 8.1), according to the official announcement. Otherwise, the details are unknown. The radiation rate inside the unit 2 reactor was also the highest.

The unit 3 reactor exploded on March 14, and the cause of the explosion is still being debated. A very strong flash of light was seen at the beginning and several detonation sounds were heard. The smoke due to the explosion was not quite the same as that of the unit 1 reactor, as seen in https://www.youtube.com/watch?v=M-nfMIspec4. In the case of the hydrogen explosion at unit 1, the vapor spread horizontally. The smoke from the unit 3 reactor was actually gray to black, rather than white. It has been suggested that a water vapor explosion was immediately accompanied by a nuclear explosion. The TEPCo have now officially called it "detonation". The hot fuel rods caused a rapid expansion of the water added (water vapor explosion). This caused the collapse of the water shield between the fuel rods stored in the storing pool, and this may have caused a supercritical condition momentarily and, hence, led to a nuclear explosion (for a comparison between the unit 1 and unit 3 explosions, see http://www.youtube.com/watch?v=eLr5_UxhoTA). The nuclear explosion was not very extensive, but there is a sign that it may have occurred at the storing pool (see Fig. 7.3), and not the reactor itself, as seen in the some of the published photos of the storing pool of unit 3 (see for example, the video on the storage pool in the unit 3 reactor published by TEPCo: http://www.youtube.com/watch?v=izaBqBGa6ew#at=98). This is of the same kind of event as that in the Chernobyl accident. If this is, indeed, the case, then not only

Fig. 8.8 Fukushima Dai-ichi NPP, unit 3 (*left*) and unit 4 reactor

fission products but also U-238, U-235, Pu-239, and neutrons would have been released. Pu-239 (and also other Pu isotopes) has been detected around the plant and up to several tens of kilometers away, though not as high as in the Chernobyl case, and, also, neutrons were observed in the premises of the plant. Data from several of the EPA's radiation monitoring stations (in the USA) recorded sudden increases of Pu-239, U-234, and U-238 after the Fukushima accidents, particularly in Hawaii, Alaska, Guam, and California. These data suggest the release of fuel rod material itself in addition to fission products, and are consistent with the notion that a nuclear explosion, indeed, took place. This description is entirely a conjecture based on what has been observed, but much uncertainty still exists regarding the explosion at unit 3.

Unit 4 was supposed to have contained no fuel rods in the reactor, and had 1,500 or so spent fuel rods in the cooling pool. It had some kind of explosion, and the result was clearly serious, as seen in Fig. 8.8. However, what exactly happened is not known. Some people believe that it was a hydrogen explosion due to a momentary failure in the cooling of the fuel rods in the pool. The fact that fuel rods are still present in the pool and no significant damage is seen on the rods may contradict this assertion. Other than several photos of the storage pool, the situation inside the plant, particularly that of the reactor container, has not been made public (or may not be known). Unfortunately, no details of the accidents have yet been revealed, partly due to the fact that the radiation rate is too high in the reactor buildings to allow human inspections.

Many fuel rods riddled with radioactive material are still present under various conditions; melted or otherwise. As long as they are covered with cold water, no major release of radioactive material into the air is likely. But Japan is vulnerable to shaking by major earthquakes, and that might trigger another major release of radiation, particularly from the unit 4 reactor. It has been discovered upon finer scrutiny that many of the Japanese nuclear reactors are situated near to or on top

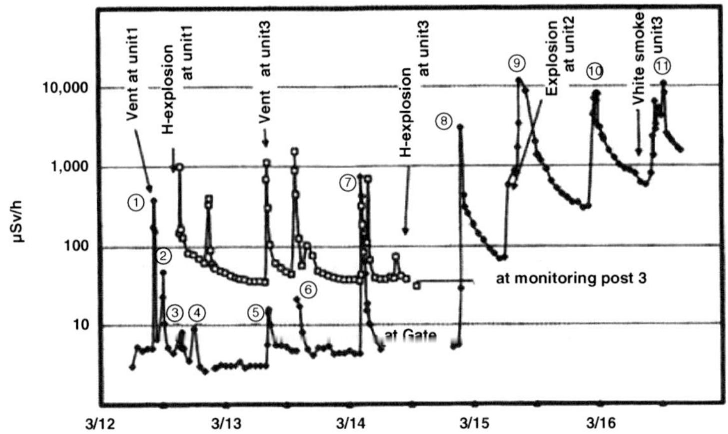

Fig. 8.9 Radiation readings on the premises of Fukushima Dai-ichi NPP (From a report by TEPCo)

of active fault lines. This makes them highly vulnerable to accidental failures caused by earthquakes; this could happen in the near future, with devastating consequences.

The amount of radiation released as the result of the accidents has been estimated by various agencies. Table 8.1 shows the data obtained by the nuclear security agency of Japan and submitted to the International Atomic Energy Agency (IAEA) in June 2011. It should be noted that significant amounts of Pu-239/240/241 were released. It was also reported so in the journal Nature (Zheng et al. 2012). Figure 8.9 shows the radiation readings at two monitoring posts on the premises of Fukushima Dai-ichi NPP. It reveals the major events during the first 5 days. The data of the radiation readings at four monitoring posts within 10 km of the facility on 2011.03.12 are shown in Fig. 8.10. These data indicate that the leakage of radioactive material had started before the vents and the explosions.

8.5.2 The Distribution of Radioactive Material Released

As shown in Table 8.1, large amounts of various radioactive materials have been released from the Fukushima Dai-ichi NPP. How, when, in what directions, and how far have they spread? Various data based on the actual measurements as well as simulations have been published. Figures 8.11, 8.12, and 8.13 show a few such data, mostly based on those reported by the Ministry of Science and Education, the Japanese government.

Figure 8.11 shows the I-131 distribution (deposited in soil) within Fukushima prefecture on samples collected and measured on June 14, 2011. As I-131 is a fast-decaying isotope, it must have been measured earlier, soon after the 3.11 accidents,

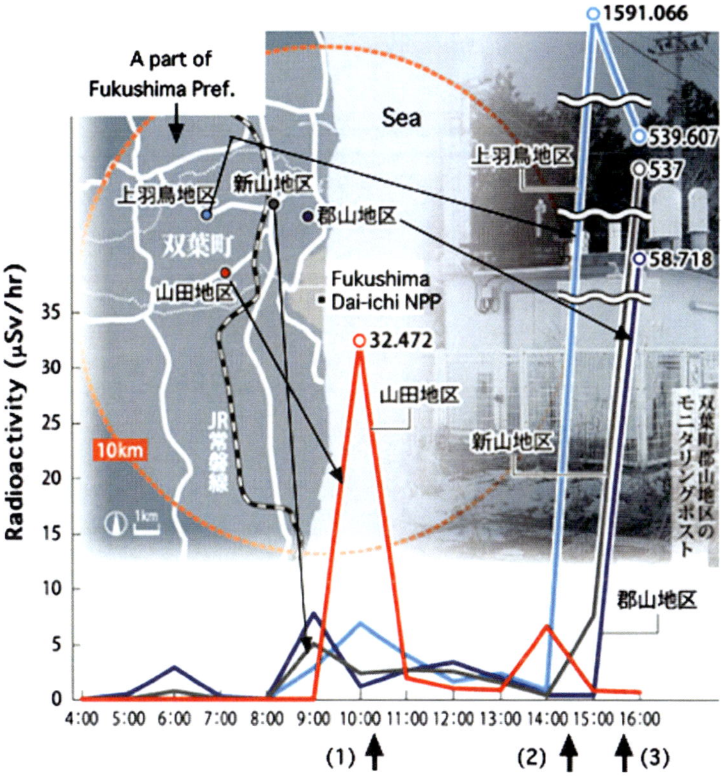

Fig. 8.10 Radiation readings at four monitoring posts within 10 km of the NPP on 2011.03.12 (*1*) first event at unit 1, (*2*) fourth event at unit 1 (*3*) hydrogen explosion at unit 1 (http://mainichi.jp/select/news/20130222k0000m040136000c.html)

but no reliable data have been published previously. This is critical, as I-131 can cause thyroid gland problems.

Figure 8.12 shows that Cs, both Cs-134 and 137, has spread far and wide, even far beyond Tokyo, which is located about 250 km southwest of the Fukushima nuclear plant. Cs-137 has a half-life of 30 years, and is one of the highest yield products in the thermonuclear fission reaction, as shown in Table 2.1. It causes various illnesses as it enters the human body, through internal irradiation with β- and γ-rays.

Figure 8.13 depicts the spread of Sr-90. Because SrO or $Sr(OH)_2$ (the likely forms of Sr released from the nuclear reactor) are less soluble and have high boiling points (compared to Cs), Sr has been thought not to spread as far and wide as Cs. The data shown in Fig. 8.13 indicate that the level of Sr is about 1/1,000th of Cs in similar locations, if the data shown in the figure are assumed to be accurate. The amount released was about 1/10th of Cs, as indicated in Table 8.1. The determination of β-emission of Sr is known to be a lot more difficult than Cs, which emits both β and γ particles.

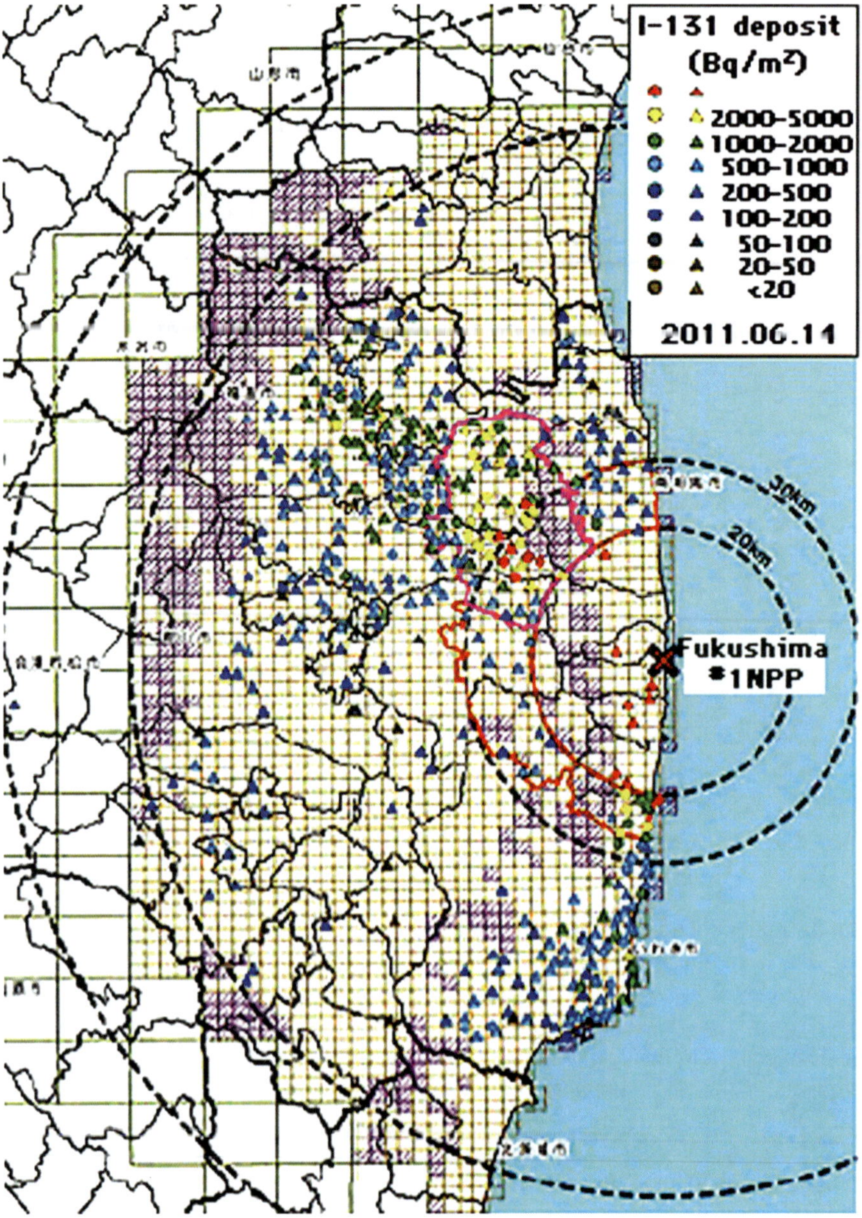

Fig. 8.11 I-131 distribution in Fukushima prefecture (Ministry of Science and Education)

These values (Bq/kg of soil) should be compared to those in previous decades in order to appreciate the increase of radiation after the Fukushima disaster. The next two figures are the change of the soil content (activity) of Cs and Sr over the last few

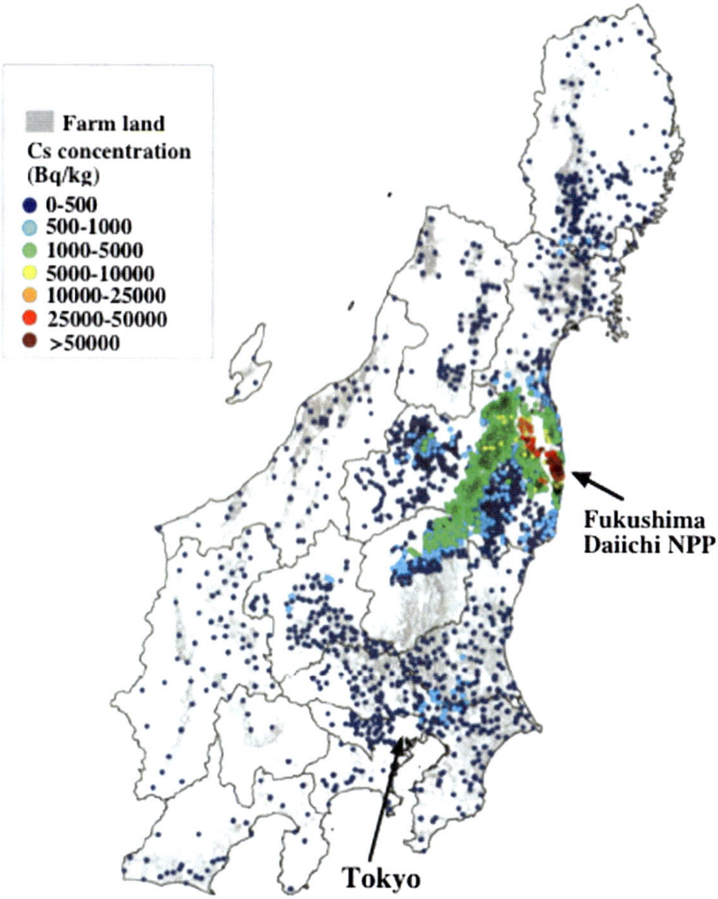

Fig. 8.12 Cs-137 deposited in soil (Ministry of Science and Education)

decades in Japan published by Japanese environmental agency (http://www.kankyo-hoshano.go.jp/study_flash.jsp?runmode=2). The Cs-137 content in soil ranges over 1–100 Bq/kg, and was gradually decreasing before the 2011.03.11 accidents. There seems to be a small sign of the Chernobyl effect, but this was not significant. In the case of Sr-90 also, it was decreasing gradually before the Fukushima disaster. The soil content of Sr was about 1/10th of Cs-137 over these decades. In view of these values, the Sr-90 figures in Fig. 8.13 after the disaster seem to be rather low in comparison with Cs (Figs. 8.14 and 8.15).

Another such data set (Fig. 8.16) is the change of Cs-137 content of Japanese daily food over the last several decades. Again, there was a declining trend over this period, but this graph shows clear indications of the effects of the Chernobyl incident. One recent estimate (http://www.asahi.com/special/10005/TKY201201180799.html) of the radiation content of the food consumed by the

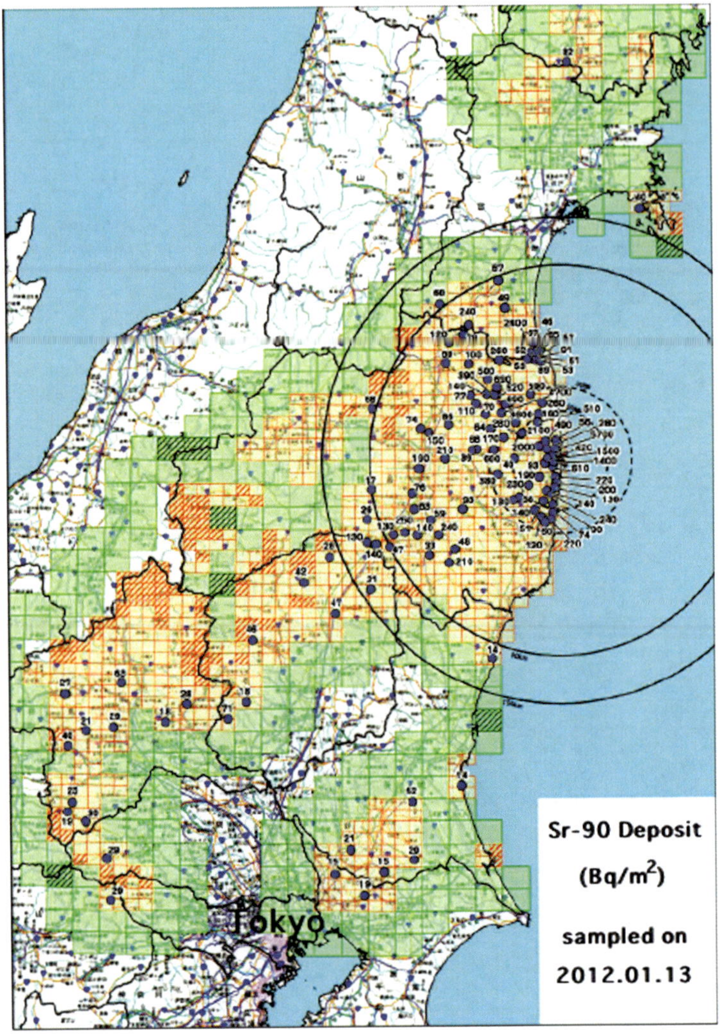

Fig. 8.13 Distribution of Sr-90 (Ministry of Science and Education)

people in Fukushima prefecture was, on average, 4 Bq/person/day, which is two orders of magnitude greater than that of the pre-Fukushima level.

A recent estimate of the distribution (October 2012) of the aerial dose (1.0 m above the ground) published by the Ministry of Science and Education is shown in Fig. 8.17. If one lives and is exposed to 1.0 μSv/h constantly over a period of a year, the dose will be 8.8 mSv/year. This is much higher than the normal allowable dose, being 1.0 mSv/y. For this allowable dose of 1.0 mSv/y to be the case, the hourly

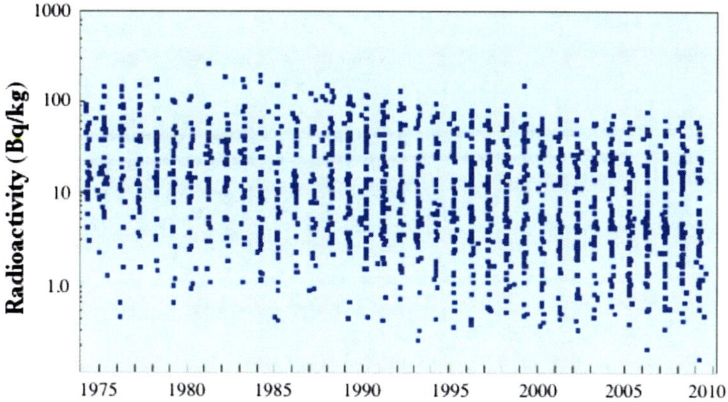

Fig. 8.14 Cs-137 content in soil in Japan before the Fukushima disaster (http://www.kankyo-hoshano.go.jp/study_flash.jsp?runmode=2)

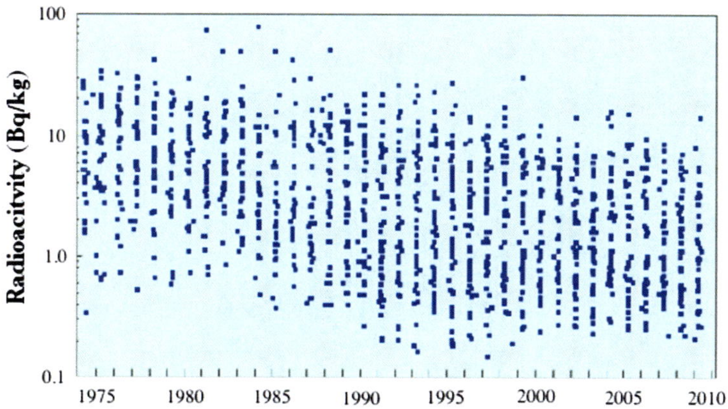

Fig. 8.15 Sr-90 content in soil in Japan before the Fukushima disaster (http://www.kankyo-hoshano.go.jp/study_flash.jsp?runmode=2)

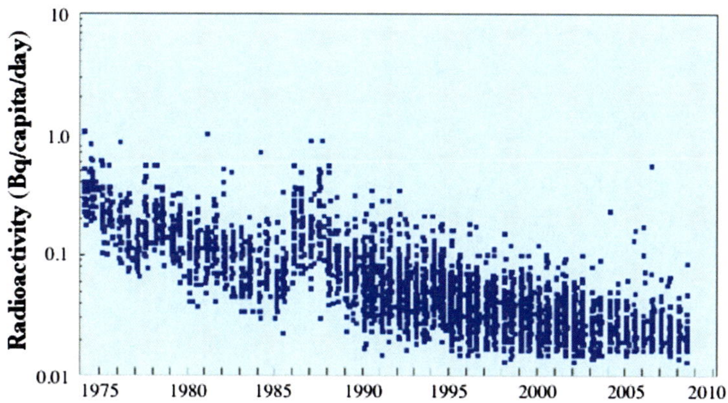

Fig. 8.16 Cs-137 content in average Japanese food per person before the Fukushima disaster (http://www.kankyo-hoshano.go.jp/study_flash.jsp?runmode=2)

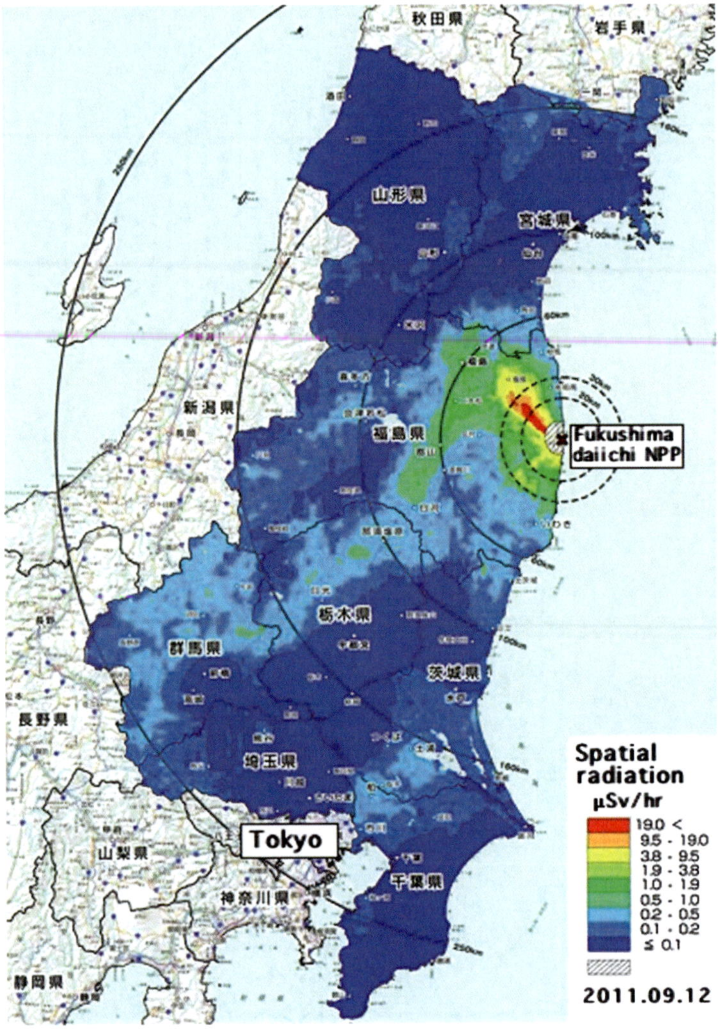

Fig. 8.17 Distribution of the spatial dose (μSv/h) of radiation (http://radioactivity.nsr.go.jp/ja/)

dose should be below 0.1 μSv/h (assuming that one is exposed over 24 h and 365 days). In any case, this figure implies that people over a large area, particularly within 100 km, would be exposed to high doses.

References

Caldicott H (2011) Nuclear power is not the answer. The New Press, New York

Chugoku Newspaper (1992) Exposure: victims of radiation speak out. Kodansha International, Tokyo/New York/London

Zheng J, Tagami K, Watanabe Y, Uchida S, Aono T, Ishii N, Yoshida S, Kubota Y, Fuma S, Ihara S (2012) Isotopic evidence of plutonium release into the environment from the Faukushima dai-ichi NPP accident. Scientific Report 2 (Nature), article #304

Part IV
The Interactions of Radiation with the Chemical/Biological World: The Biohazards of Radiation

Part IV is the main theme of this treatise. Radiation effects on chemical/biological systems are discussed in terms of interactions of radioactive particles (including photons) with atoms and molecules. The main effect is typically designated as "ionization", i.e., kicking out an electron from a chemical entity, resulting in the formation of an ion or the cleavage of a chemical bond to form free radical(s). A single radioactive particle could ionize a large number of chemical entities, as its kinetic energy is several thousand to more than a million times as large as the typical chemical energy required for ionization to occur or to break bonds. The distinction between "external" and "internal" exposure is emphasized because it is very important to differentiate them in order to critically evaluate the radiation effects on biological systems.

Chapter 9
Interactions of Radiation with the Chemical World

9.1 Energy Changes in the Chemical World

The chemical world consists of atoms and assemblies of atoms (through chemical bonds), i.e., chemical compounds formed through ionic, covalent, and intermetallic bonds. Chemical compounds can change as a result of the removal of electrons, and the breakage and formation of chemical bonds. In ordinary chemical reactions, these processes are carried out chemically, i.e., by some other chemical means. For example, the removal of an electron from a chemical entity is accomplished by an agent that has a strong tendency to acquiring an electron; such an agent is called an "oxidant". In other words, the removal of an electron(s) from a chemical entity is known as "oxidation". The reverse process is termed "reduction". A chemical bond can be severed in a number of ways. For example, the $H_3C–H$ bond can be broken by another agent that binds strongly with "H", i.e., $H_3C–H + X \rightarrow H_3C^{\bullet} + H–X$. In all chemical processes, the only thing that changes takes place at the level of "electrons", which interact with each other and nuclei through the electromagnetic force.

The energy of the electron in a hydrogen atom is -13.6 eV in its lowest level (closest to the nucleus). The negative sign indicates that the energy of that electron is much lower (more stable) than the situation where the electron is free and not associated with the nucleus (this energy is taken as "zero"). The lowest energy in an N atom is -427 eV. As the atomic number becomes larger, this energy (magnitude) also becomes larger. In ordinary chemical reactions, however, the electrons in lower energy levels are not involved because they are difficult to move, and, usually, only the electrons which are most readily movable are involved in chemical change/chemical reactions. These electrons are called "valence electrons". The typical chemical energy changes range from 0.1 to 100 eV at the most.

The radiation emitted from the nuclear decay process has a characteristic energy in the form of the kinetic energy of a particle or electromagnetic energy associated with a photon. The α-particle is the helium nucleus with two protons and two neutrons. As discussed earlier, an α-particle is a very stable entity, and its emission

E. Ochiai, *Hiroshima to Fukushima*, Science Policy Reports,
DOI 10.1007/978-3-642-38727-2_9, © Springer-Verlag Berlin Heidelberg 2014

relieves the repulsive tension within a heavy nucleus. Hence, the decay energy or α-particle energy is relatively large, typically of about 5 MeV, as shown in Table 2.1. It is heavy as a particle and carries two positive electric charges.

As neutrino emission accompanies a β-decay, a β-particle has somewhat variable energy, though it is distributed around an average. A β-particle is an electron of high speed and much less massive, about 1/7,200th, than an α-particle, and carries a negative charge. β-emission occurs when the nuclide contains too many neutrons to be stable. In this case, a neutron is converted to a proton, emitting an electron; ${}_0n^1 \rightarrow {}_1p^1 + {}_{-1}e^0 + \nu'$ (antineutrino). The energy of β-particles varies over the range of 10 keV to about 5 MeV, but those of most nuclides range from around one to several hundred keV. The antielectron, positron β^+, is another particle emitted from a decaying nucleus, where protons are present in excess (${}_1p^1 \rightarrow {}_0n^1 + {}_{+1}e^0$ (β^+) + ν (neutrino)). Such a nucleus can also decay by capturing an electron: ${}_1p^1 + {}_{-1}e^0 \rightarrow {}_0n^1$ (EC).

A γ-ray is an electromagnetic wave, which covers a wide range of radiations, and is characterized by wavelength λ and frequency ν. They are related in the form of $c = \nu\lambda$, where c is the speed of light (electromagnetic wave in vacuum). The FM radio frequency familiar to us is something on the order of 70 MHz, which is $\nu = 70 \times 10^6\ s^{-1}$. The common microwave oven and relatively long-distance electronic communication use microwaves, whose wavelength ranges over 1 mm to 1 m. Let's calculate the wavelength and frequency of these two electromagnetic waves: radio frequency 70 MHz corresponds to *ca.* 4.3 m in wavelength, whereas the microwave of 1 cm wavelength corresponds to $\nu = 3 \times 10^{10}\ s^{-1}$. The visible range of light (electromagnetic wave) is from about 800 nm (red) to 350 nm (violet) in wavelength; 500 nm light has a frequency of $63 \times 10^{14}\ s^{-1}$. Light longer than the visible range is called infrared light, and it provides heat. Shorter than visible light is ultraviolet light; we see that 100 nm corresponds to $3 \times 10^{15}\ s^{-1}$.

When light (electromagnetic wave) interacts with a chemical, it behaves as a particle termed a "photon", whose energy is proportional to its frequency, i.e., E (of photon) = $h\nu$, where h is Planck's constant and ν is the frequency. Let's then calculate the energy of the typical electromagnetic waves mentioned above. The FM radio 70 MHz's energy is 3×10^{-7} eV; 500 nm visible light is 2.5 eV; and 100 nm ultraviolet light is 12.5 eV.

Now, let's look at X-rays and γ-rays or, rather, their photons. A typical γ-ray of Cs-137 has an energy of 662 keV. A γ-ray typically ranges from 10 to 5,000 keV. An X-ray is created artificially by bombarding a high-speed electron beam onto a metal, such as tungsten or copper. The wavelength of commonly used X-rays ranges from 10^{-2} to 10 nm, i.e., 100 eV ($3 \times 10^{16}\ s^{-1}$) to 100 keV ($3 \times 10^{19}\ s^{-1}$). Hence, X-rays and γ-rays are very similar, and the latter is about an order or two stronger than a typical X-ray. It must be pointed out, though, that a γ-ray is based on nuclear decay (energy transition within the nucleus), but an X-ray is produced by electronic transition between high electronic states to lower (mostly the lowest K) levels. Energy-wise, these two processes partially overlap.

An energy comparison among many kinds of radiation is given in Fig. 9.1. The range of energy involved in chemical changes is shown in blue, and the energy of

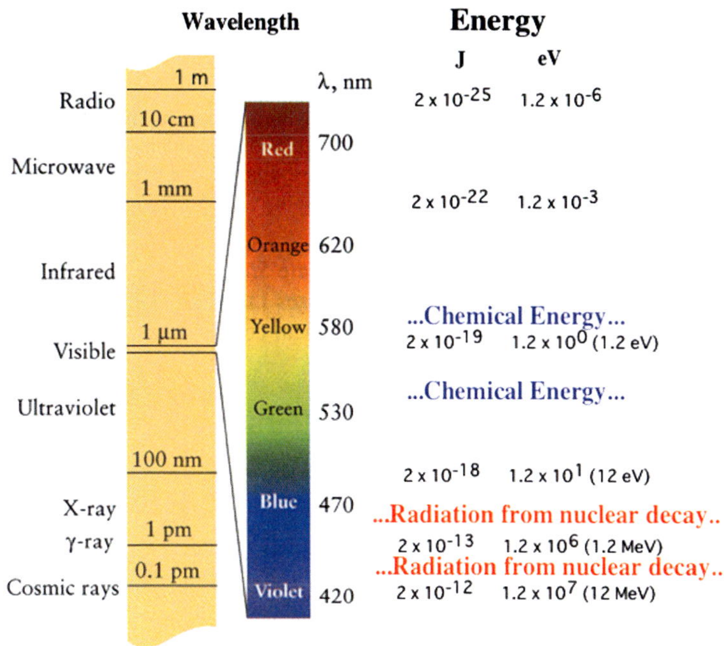

Fig. 9.1 The energy ranges of various radiations

radiation from nuclear decay is indicated in red. There are several orders of magnitude of difference between them, up to several millions. The essence of radiation effects on the chemical world, including life systems, is the interaction (collisions, assaults) of radiation particles of very high energy upon it.

9.2 Interaction of High-Energy Radiation with Chemicals

9.2.1 Photons (X-rays and γ-rays)

Let us first look at photons, which have no mass and no electric charge. That is the action of X-rays or γ-rays. They can collide with loosely bound electron(s) in a chemical compound or they may collide with an electron in the inner core or even the nucleus. The Compton effect is predominant for photons of medium energy, 0.3–3 MeV. With photons of lower energy than 0.1 MeV or so, the photoelectric effect is the most important, while photons of higher energy (>5 MeV) lead to electron–positron pair formation.

When the photon energy is not very high, it is absorbed by the system and the absorbed energy is transferred to an electron that will escape from its orbital. The kinetic energy of the ejected electron is the difference between the energy of

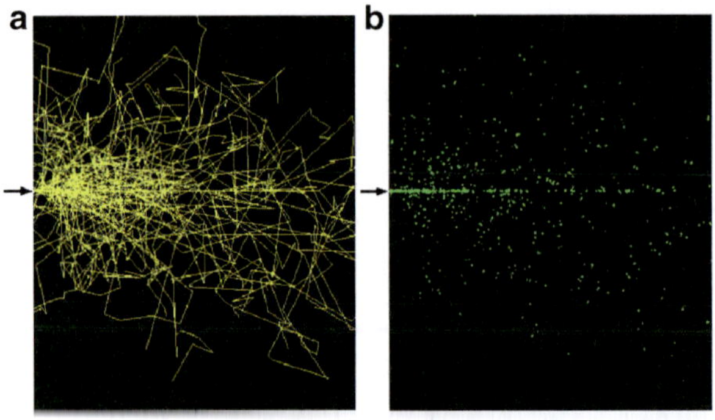

Fig. 9.2 (**a**) The trace of photons (100 photons of 1 MeV) injected into a medium of carbon with 20 cm radius and 30 cm depth, and (**b**) electrons formed (Cited from http://www-egs.slac.stanford.edu/cgi-bin/egsdemo/)

the incoming photon and the energy required to eject the electron. The bombarded entity has lost an electron, thus, it is ionized. The energy required to form an ionized entity is typically 25–40 eV, which is higher than the ionization energy, and the difference is used for excitation. The ejected electron can then act as a radiation particle, if it carries enough energy.

With a higher energy, the photon not only ejects an electron, but it also still retains a sufficient energy and is scattered to travel in a new direction. This is the Compton effect.

The scattered photon may interact again to eject another electron, and so on, until it is absorbed in the manner of photoelectric absorption. This effect will produce a number of ionized entities, as well as secondary high-speed electrons, which could act as β-particles.

Photons with energy higher than 5 MeV (though 1.02 MeV is enough theoretically) lead to the formation of an electron–positron pair, each of which departs in an opposite direction to the other.

A simulation has been performed that calculates how 100 photons of energy 1 MeV would interact with a chemical substance (carbon atom ensemble of 20 cm radius and 30 cm depth). The trace of photons and electrons produced is shown in Fig. 9.2a, b.

Photons themselves travel long distances, but the electrons formed travel only short distances, as seen in this figure. Some photons themselves are seen to traverse this medium and come out the other side to a significant degree (Fig. 9.2a). X-rays and γ-rays can travel up to 1 m in soft tissues, and, indeed, penetrate the human body (20 cm or so thick). That is why X-rays can be used as a diagnostic tool, and the measurement of γ-rays coming out of the whole body may give some measure of the internal exposure of radiation (discussed later). X-rays and γ-rays traverse a long distance, typically up to 100 m in air. They can be stopped by a thick material of heavy elements, such as lead or concrete.

9.2.2 *Beta (β)-Particles (Electrons) and β⁺-Particles (Positrons)*

A β-particle is a high-speed (close to the light speed) electron and carries a negative charge. An atom or molecule is essentially an electron cloud with a tiny positively charged nucleus inside. An electron may be scattered by an atom (or molecule), and it may or may not lose its kinetic energy, changing only its direction of flight (Rutherford scattering). The former case is that scattering is inelastic, meaning that a part of the energy is turned into an electromagnetic wave (X-ray) upon deflection (this phenomenon is termed as "Bremsstrahlung").

A more likely phenomenon is that the colliding electron ejects an electron from an atom or a molecule, thus, ionizing it, and then moves on. The β-particle loses its energy somewhat (something like up to 50 eV, depending on what kinds of ionized entities are formed). The ejected electron may behave like a β-particle if it has enough energy, or it may attach to a chemical entity to turn it into a negatively charged species. The β-particle will continue in this way as it travels, until no sufficient kinetic energy is left.

A β-particle may eject an electron from an inner orbital (of an atom or molecule). When the vacancy is filled by the dropping of an electron from an outer orbital, the energy difference would be released as a photon (e.g., in the range of an X-ray or otherwise). A β-particle may also excite an electron from a lower state to a higher state, and its reversal could then emit ultraviolet light.

How far a β-particle will travel depends on its kinetic energy and the medium through which it is passing. It is estimated that a β-particle with energy 2 MeV will travel up to about 10 m in air, but up to 10 mm in water, and, likely, several tenths of millimeters in human tissue. Likely, it will ionize 250 or so molecules in its path. Most β-particles will be stopped by thin aluminum foil or a wood plate of thickness 1 cm. That is, it travels only less than 1 mm in a metal like aluminum, where the atoms are packed. The slowing effect on moving electrons is stronger with heavier elements.

β-emission occurs typically with neutron-rich isotopes (nuclei), while a proton-rich nucleus tends to emit a positron (β^+). The behavior of a positron is similar to an electron while it is travelling at a very high speed. However, as it slows down, it tends to be captured by an electron. When matter (electrons) and antimatter (positrons) combine, they dematerialize; that is, they turn into energy, and a photon forms, i.e., γ-ray.

9.2.3 *Alpha (α)-Particles*

An α-particle is equivalent to a helium nucleus, but at a high speed. It consists of two protons and two neutrons, and is heavy (by 7,200 times compared an electron) and carries two positive electric charges (Sect. 2.4). The fact that an α-ray is scattered (only occasionally) by a gold foil led Rutherford to arrive at the current model of an atom, in which a positively charged very small nucleus is surrounded by an electron cloud; this effect is called Rutherford scattering, and is caused by the Coulomb repulsion between the nucleus and the α-particle.

It might impact an atom nonelastically, and, hence, some energy is given to the photon emitted (Bremsstrahlung). More importantly, the impact of this heavy and electrically charged particle is very strong, and it ejects an electron(s) from the impacted entity, i.e., ionizes it or breaks chemical bond(s). As it is heavy and its kinetic momentum is high, it tends to proceed in a straight line. During this process, the α-particle loses some energy. As it proceeds, it may acquire electron(s) and eventually turn into a slow-moving helium atom. At this stage, the radiation effect is extinguished. An α-particle, though its impact is strong, does not travel very far; that is, it is not very penetrating. An α-particle of 5 MeV will travel only about 4 cm in air, at most only about 100 μm in water, and 4 μm or so in mammalian tissues, and will not penetrate a sheet of paper. However, it can ionize as many as several thousands of molecules over a path of 1 μm. Typical tracks of α- and β-particles in water are shown in Fig. 9.3 (adapted from Wikipedia).

9.2.4 Neutrons

A neutron, which does not have an electric charge, can penetrate well the chemical world. It can approach a nucleus and will be scattered by it. But if it has enough energy, it may eject a proton, called a "recoil proton". This proton will then behave like a lighter α-particle, until it acquires an electron and turns into a hydrogen atom. When it slows down, it can be absorbed by a nucleus. The nucleus with an extra neutron can become radioactive, likely, a β-emitter. Neutrons, thus, create radioactive material inside the chemical world. As it carries no electric charge, it travels a long distance: up to 100 m in air and up to about 1 m in water.

The major ways of interaction of various radiation particles with the chemical world (nuclei and electrons) are summarized in Table 9.1.

9.2.5 Radiation on Nonliving Chemicals

Let's see how much effect radiation would have on a chemical which it impacts. Let's assume that the radiation impact (dose) is 1 Gy (1 J/kg). We will see the effect only in terms of energy. That is, 1 J of radiation energy is absorbed by 1 kg of a material. This is an extremely small amount of energy; if the heat capacity of the material is 1 deg/cal/g (as with water), then 1 J will raise the temperature of the material by only 0.00024 deg. This would cause no substantial change in the material.

Next, let's see its effect in terms of radiation particles. Suppose this radiation carries 1 MeV of energy per particle (α, β, or γ). One joule is equivalent to 6.24×10^{18} eV. Therefore, a 1 J impact is given by about 6×10^{12} radiation

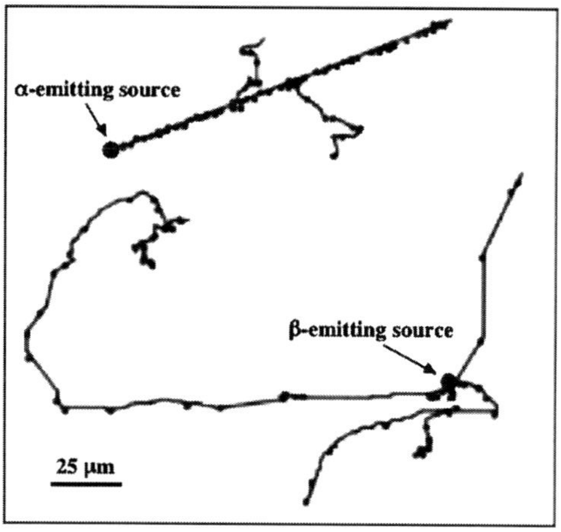

Fig. 9.3 Tracks of α- and β-particles in water (Modified from Wikipedia)

Table 9.1 Types of interactions of ionizing radiation with matter

Incident radiation	In collision with	Elastic collision	Inelastic collision	Absorption
Photons (X-ray, γ-ray)	Nucleus	Thompson Sctrg	Mossbauer effect	Photodisintegration
	Electron	Rayleigh Sctrg	Compton effect	Photoelectric effect
	Field	Delbrück Sctrg		Pair production
β^- or β^+	Nucleus	Rutherford Sctrg	Bremsstrahlung	Electron capture
	Electron		Ionization/ excitation	Annihilation (β^+)
α	Nucleus	Rutherford Sctrg	Bremsstrahlung	Transmutation
	Electron		Ionization/ excitation	
Neutron	Nucleus	Recoil	Resonance scattering	Radioactivation and nuclear reactions

Sctrg scattering

particles. Each particle is assumed to ionize or otherwise affect 100–1,000 entities as it moves through the chemical material. That is, 1 Gy of radiation will impact on about 6×10^{15} chemical entities (particles). As seen in Sect. 3.2, 1 kg of a chemical material typically contains some 10^{24} atoms or molecules. Therefore, only about one in 0.1 billion atoms or molecules will be impacted by the 1 Gy radiation dose. This may cause minor disorders that would not affect the overall chemical material. Soon, the affected entities would return to normal states by exchanging electrons or otherwise, releasing some energy, which raises the temperature of the body by

0.000024 deg, as calculated above. This much exposure very likely has no appreciable effect on inanimate chemical material.

It can be inferred that 1,000 Gy or more may disrupt the integrity of the material somewhat, as about one (or more) in 0.1 million atoms or molecules is affected. The overall effects may depend on the material that absorbs the radiation. A structurally fragile material may be affected significantly. M. S. Curie observed that a glass tube containing several milligrams of radium darkened within a few months and cracked.

Some chemical compounds are decomposed; more than 10 kGy is said to be required in order to give an appreciable effect (loss of elasticity and increase of brittleness) on metals (Chap. 7 in Choppin et al. (2002)). If the dose is lower than this extreme, most of the material may simply be raised in temperature by 0.24 deg or more, and expanded a little bit, but healed eventually to return to its original state. However, many chemical substances may not be healed to their original state. One such example is polyvinylchloride (PVC). It is well known that PVC becomes brittle upon exposure to sunlight. Chlorine is readily cut off from PVC by ultraviolet light, and more so by γ-radiation. Then, the PVC would be attacked by dioxygen and its chain structure would be broken, resulting in a brittle state. Apparently, this is now happening at the accident site of Fukushima Dai-ichi NPP (http://aoitombo.s100.xrea.com/enbitube.html). This website (on 2012.10.20) reports on several incidents at the NPP site. Workers stepped on PVC hoses which broke, resulting in the leakage of contaminated water. The causes of this could be the long-term exposure to sunlight, but it is highly unlikely that it was the major cause. The degradation of the PVC hoses is likely facilitated by the radiation at the NPP.

In the mantle, radiation (from U-238, Th-232, and its decay products, such as Ra, K-40, etc.) affects the rock material, disrupts the integrity of the structure, and melts it, as the radiation energy is converted to heat.

Reference

Choppin RG, Liljenzin J-O, Rydberg J (2002) Radiochemistry and nuclear chemistry. Butterworth-Heinemann, Woburn

Chapter 10
Chemicals and Their Behaviors in Biological Systems

10.1 A Survey

About 30–35 chemical elements are required in living organisms (Fig. 3.1). In addition, a living body may contain all naturally occurring isotopes. Nonessential elements may get into a living body inadvertently, as it does not have a very effective means to reject unnecessary entities while incorporating the necessary ones. There is some correlation between the concentration in the human body and that in its surroundings. Figure 10.1 shows such a correlation with the concentration in the ocean. Some of these nonessential elements are tolerated by the body, as they are not extremely toxic. However, others can be toxic and dealt with by the body, although often not sufficiently, so that toxic effects manifest beyond the safe threshold (Chaps. 16 and 17 in Ochiai (2011)).

The major essential elements H, C, N, O, S, and P constitute the organic compounds. Na and K are cations that balance the osmotic pressure inside and outside cells, and occasional sudden reversals of their concentrations create electrical signals; this is the basis of the transmission of nerve signals. Ca is present mostly in the hard tissues of the human body, i.e., teeth and bones, but distribute widely throughout the body, playing a variety of essential physiological roles. Mg is also universally required by all bodily functions, and Cl is the most important counteranion and is universally distributed in the human body (Ochiai 1977, 2008).

Fe and Zn play essential functions in many widely distributed proteins and enzymes. Other essential elements are more or less specific and concentrated in specific tissues/organs. One such important element is iodine. Iodine in the human body is a constituent of the thyroid hormone, thyroxine, and is concentrated in the thyroid gland.

The human body and all living organisms have developed mechanisms to adequately deal with these essential elements over the course of evolution. As the biological systems can do this only with chemical means, which are not very specific in certain cases, they may not be able to adequately distinguish between two certain similar elements. For example, P is absorbed and bound to

E. Ochiai, *Hiroshima to Fukushima*, Science Policy Reports,
DOI 10.1007/978-3-642-38727-2_10,

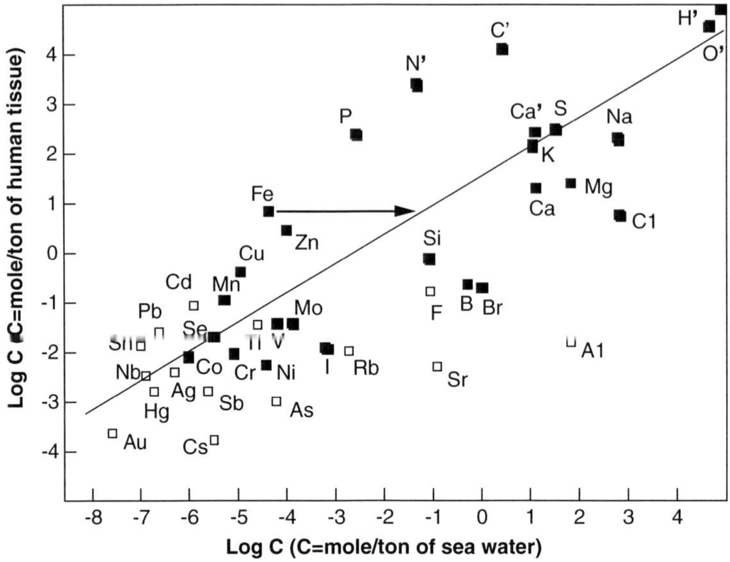

Fig. 10.1 Correlation of the elemental composition between the human body and current seawater; the *black dots* are essential and the *white dots* are nonessential (the *arrow* indicates the concentration of Fe in the ancient sea; adapted from Ochiai 2008)

biocompounds usually in the form of PO_4^{3-} (phosphate). AsO_4^{3-} (arsenate) is very similar to PO_4^{3-} in size and chemistry, and, hence, the body may mistake AsO_4^{3-} for PO_4^{3-}. Once AsO_4^{3-} is incorporated into an important compound, it may manifest its toxic effect, because the chemical similarity is not sufficient for it to mimic PO_4^{3-} in terms of bioactivities.

The metabolisms of most of the essential elements have been well understood. Biochemistry essentially aims to describe how C, H, N, O, S, and P behave in living systems. The so-called "bioinorganic chemistry" deals with the behaviors of other elements, such as Ca, Zn, Fe, Cu, Mo, Mn, Co, Hg, Cd, As, etc. (Ochiai 2008), although not all of them have yet been studied in full.

It is not the intention here to explore all these mechanisms in detail. However, a few points will be mentioned, which might be relevant to the discussion of radiation effects by a few important elements.

10.2 Potassium and Cesium

The elements in the first column in the periodic chart (Fig. 1.3) are called "alkali metals". They are, from the top, Li, Na, K, Rb, Cs, and Fr. The major naturally occurring isotopes of these elements are stable and nonradioactive, except for Fr (francium), which is an α-emitter. Only K has a naturally occurring radioisotope,

K-40. Of these elements, only Na and K are essential to all living organisms, though Li, Rb, and Cs are present in minute quantities in the human body. Besides, Li compounds are used as a medicine for some mental illnesses, such as depression. Thus, stable Li, Rb, and Cs seem to be tolerated at very low levels. In the human body, Rb is present at 9 ppm (parts per million), Cs at 20 ppb (parts per billion), and Li at 30 ppb. All of these elements are very electropositive, having low ionization potentials, and exist as (+1) cations in the aerobic environment and biological systems. All these (+1) cations behave similarly in chemical terms.

However, biological systems need to distinguish between some of them. How they do so is not simple, and the degree of discrimination is sometimes not quite strict, because there are not any very good chemical factors on which distinction can be made. The size of (+1) cations is one of such factors. They are: Li 58 pm, Na 102 pm, K 138 pm, Rb 149 pm, and Cs 170 pm. Another relevant factor is the hydration energy of the cation. They are: Li 543 KJ/mol, Na 433 KJ/mol, K 349 KJ/mol, Rb 333 KJ/mol, and Cs 290 KJ/mol. Organisms have devised a number of mechanisms to distinguish Na(+) and K(+) utilizing these differences (Chap. 15 in Ochiai (1977)). Because of these mechanisms, the concentrations of Na(+) and K(+) are distinct in and out of cells: in blood plasma or interstitial fluid, the concentrations are Na 150 mM (10^{-3} mol/L) versus K 5 mM, whereas in the cell fluid (cytoplasm), the concentrations are Na 14 mM versus K 157 mM. Specific channels are present, each of which allows either the active (Na/K-ATPases) or passive movement of either Na(+) or K(+). Bacteria produce a number of compounds termed “ionophores”. They rely mainly on the size difference, i.e., one has a specific binding site of appropriate size for K(+)but not Na(+) (Chap. 15 in Ochiai (1977)). However, many of them have less than ideal specificity. Therefore, one may capture both Na(+) and K(+), though the capture ratio may not be the same. A few examples of binding constants of ionophores for cations are given below (Chap. 15 in Ochiai (1977)):

(a) Nonactin: Na 2.1×10^2, K 3.9×10^3, Rb 3.3×10^3, Cs 7.3×10^2
(b) Valinomycin: Na 3.7×10^0, K 6.3×10^4, Rb 1.4×10^5, Cs 3.1×10^4

These data suggest that, for example, Cs(+1) should behave chemically similarly to K(+1), but that in the human body (or other organisms), Cs might be discriminated against K. For example, Cs(+) may not pass a K-channel as smoothly as K(+), because it is much larger than K(+), as shown below.

The behavior of Cs seems to have attracted attention since the introduction of radioactive Cs via the nuclear industry. One of the most comprehensive studies on Cs in the human body (Leggett et al. 2003) indicates the following:

(1) In terms of Cs concentration in tissues/organs, the skeletal muscle has the highest Cs concentration under physiological conditions. This concentration is about 2.5 times greater than that found in the heart, kidney, gastrointestinal tissues, liver, lung, spleen, and pancreas; in these organs/tissues, the Cs levels are about the same. The other organs/tissues have about one order of magnitude smaller Cs concentrations. The blood plasma content is about 1/50th of that in

muscle. In other words, Cs is mostly absorbed in tissues/organs. These numbers are those under steady-state physiological conditions.

(2) How well a chemical species is transferred from blood plasma into tissue is defined as the extraction fraction $E_{T,\ X}$, where T is the tissue and X is the element. From a number of studies on experimental animals, E values for K and Rb have been estimated to be 0.60–0.90 and 0.4–0.85, respectively, in all tissues, except for the brain. The value for the brain was estimated to be 0.015 for K and 0.01 for Rb. An estimate for heart muscle in dogs indicated $E_{T,X} = 0.71$ (range 0.64–0.80) for K, 0.65 (0.58–0.76) for Rb, and 0.22 (0.09–0.30) for Cs.

(3) The transport of Cs through K-channels is sluggish compared with the transport of K and Rb; mostly, the ratio Rb/K is near 0.7, and that of Cs/K is 0.02–0.2. In the transport through Na pump (Na/K-ATPase), Cs/K = 0.25 and Cs/Rb = 0.3.

(4) For the kidney, $E_K = 0.9$, $E_{Rb} = 0.85$, and $E_{Cs} = 0.2$. That is, filtering at the kidney is only about 20 % efficient with Cs(+) as compared to K(+) and Rb(+).

(5) Using the model proposed in this paper, the authors estimated the time change of the Cs concentration in several tissues after a single ingestion of Cs, which is shown in Fig. 10.2. In most tissues, the Cs concentration diminishes relatively quickly but then slows down, tending to become steady, but the skeletal muscle accumulates Cs, reaching a maximum after about 10 days, and then slowly decreases. The decrease of the overall Cs level in the body does not follow an exponential decay curve, as often assumed. The overall half-life (only in the sense that about half of the original dose is removed from the body) seems to be several tens of days, as presumed by the ICRP document. It diminishes quickly at first, but does not diminish as fast as that predicted by the exponential curve and tends to stagnate, as seen in Fig. 10.2. This has significant implications in the internal radiation effects by radioactive Cs.

Figure 10.2 is a simulation calculation based on a model and a set of parameters, but the reality may not be exactly like this. This is a single-shot process, and applies to a situation where people consume normally uncontaminated foods and drinks (containing naturally occurring Cs). If the Cs ingested is uncontaminated with radioactive isotopes, the chemical toxicity of nonradioactive Cs is usually low enough for it to be tolerated. In reality, however, people today may eat contaminated foods every day or occasionally over the course of several days; the result seems to be the accumulation of Cs in certain organs/tissues, as the body would not distinguish the radioactive from the nonradioactive Cs. Figure 10.2 suggests that Cs would be accumulated the most in skeletal muscle, followed by the kidney, liver, heart, and the brain. We will discuss briefly how food or drink contaminated with radioactive Cs might result in significant internal radiation exposure.

A team at Tohoku University and others recently published a study on the distribution of radioactive substances in cattle exposed to the radiation from Fukushima Dai-ichi NPP (Fukuda et al. 2013). They found the highest

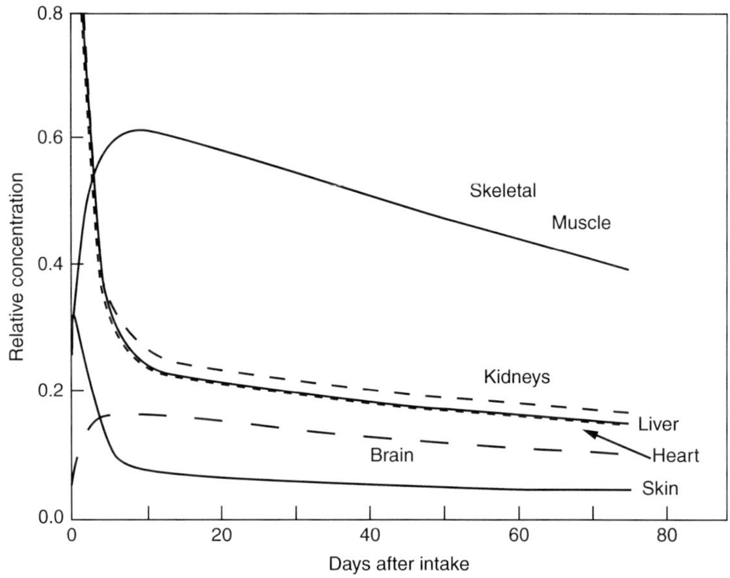

Fig. 10.2 Change in Cs concentrations in organs after an intake of Cs (Fig. 5 in Leggett et al. 2003)

concentration of Cs-137 in skeletal muscle, followed by the urinary bladder, kidney, heart, lung, liver, and the thyroid gland. This result is consistent with Fig. 10.2. The researchers also found that the Cs concentration in fetuses is about 20 % and that in calves is about 50 % higher than that in the mothers. These results are also consistent with the findings on the Cs distributions among the various organs of the dead bodies following the Chernobyl incident (Fig. 14.2). They also determined the distribution of radioactive Ag-111m and Te-129m.

An experimental study about Cs distribution was conducted by injecting a single dose of Cs-137 into pregnant mice on day 17 of gestation (Matsusaka et al. 1997). The concentration of Cs-137 was the highest in the salivary gland and kidney, followed by the heart and then muscle, and the concentration in fetuses was not high. It was pointed out that the salivary gland is close to the thyroid gland. It has been observed that Cs would not often cross the placenta barrier.

It must be pointed out, again, that the biological system, as it operates in chemical terms, would not distinguish between isotopes, i.e., stable K-39 and K-41, and unstable (radioactive) K-40 behave in the same manner; stable C-12 and C-13, and radioactive C-14 do so too (see the note at the end of this section). The same applies to any chemical elements, including I-127 versus I-131 (and other radioactive isotopes of I), or Cs-134, 135, 137, and others (refer to Table 2.1). In the current discussion of the health effects of radiation, only a single isotope is stressed and others are mostly ignored. For example, I-131 is the one that people are the most concerned with, and it has a short half-life (8 days). However, the thyroid gland cannot distinguish among isotopes of iodine, and, hence, it can incorporate not only I-131 but also I-129, which has a much longer half-life of 15.7 million years. However, I-129 is much less radioactive, being about one billionth of the same quantity of I-131. Likewise, Cs-135 has a much longer half-life of 2.3 million years compared to the currently concerned Cs-137. How these longer-lasting isotopes would impact on the slow-developing health effects needs to be investigated.

10.3 Calcium and Strontium

The second column of the periodic chart lists Be, Mg, Ca, Sr, Ba, and Ra. Mg and Ca are essential for all organisms. The normal concentrations of these elements in the human body are: Mg 410 ppm (by weight), Ca 14,000 ppm, Sr 4 ppm, and Ba 0.2 ppm. Since there is no naturally occurring radioactive Sr, this Sr in the human body is ideally entirely stable Sr (mostly Sr-88 and a small amount of Sr-86, 87). It is present mostly in the bones and teeth, very likely as phosphate, just like calcium phosphate. Sr is absorbed through the mechanisms to absorb Ca, because Sr is close to Ca in many ways. There are many Ca(+2) absorption mechanisms (see Chap. 10 in Ochiai, 2008), including Ca-ATPase. Both Ca and Sr are (+2) cations under normal conditions, and their sizes are also similar. The ionic radii of these cations (+2) are: Be 27 pm, Mg 72, Ca 100, Sr 116, and Ba 136 pm.

Though the majority of Ca is in bones and teeth, Ca(+2) is also widely distributed in the whole body at low levels, and plays, as a second messenger, essential physiological roles in muscle contraction, neurotransmitter release and other secretion processes, cell proliferation, and others. There are widespread and elaborate mechanisms to deal with Ca(+2) in the whole body. Sr(+2) very likely replaces some Ca(+2) in proportion to their concentration ratio, and is tolerated usually because of its low level of Sr(+2) and its stability (nonradioactive). When radioactive Sr-89/90 replaces Ca(+2) in proportion to the ratio of respective concentrations, the radioactivity would exert damaging effects in the areas occupied by the radioactive Sr. Of course, the majority of Sr would be in the bones and teeth, and, hence, it affects the bone marrow, but it is likely that other tissues and organs that require Ca may be affected by the radioactive Sr.

By the way, many marine crustaceans use calcium carbonate as their outer shell, while very few of them use strontium sulfate instead. Sr is known to be essential only for these creatures. Barium (Ba) and then radium (Ra) are located below Sr in the periodic chart, and, hence, they might behave similarly to Ca. However, they are significantly different from Ca in terms of size and chemical reactivities, and radioactive Ba and Ra would distribute in the body differently from Ca. However, it is likely to distribute mostly in the bones and teeth.

10.4 Other Elements

There are many other elements present in the human body; some of them are essential but others are not and, yet, are incorporated because of their presence in the environment (Fig. 10.1). A few elements have been studied fairly extensively, but others have not (Ochiai 2008); examples include iron (Fe), zinc (Zn), and copper (Cu). These are essential elements, but no counterparts of the radioactive isotopes would show up in humans under ordinary circumstances.

The elements or, rather, radioisotopes produced in uranium fission reactions are given in Table 2.1; all of them except for iodine, potassium, and molybdenum are nonessential and rather toxic. A few of the elements from Table 2.1 will be mentioned below.

Molybdenum (Mo) is essential for several specific enzymes, and is dealt with properly in living organisms. The average content of Mo in the human body is 70 ppb. High concentrations are found in the liver and kidney, and lower concentrations are observed in the vertebrates. The liver is where many Mo-containing enzymes are present. It is typically present as molybdate (MoO_4^{2-}) in aqueous media and likely in soil, and it is the form that is absorbed into the body. Many stable isotopes are known for Mo; they are as follows in the order of natural abundance: Mo-98, Mo-96, Mo-95, Mo-92, Mo-100, Mo-97, and Mo-94. The only radioisotope is Mo-99, which is a β-emitter, and likely behaves similarly to all the other stable Mo isotopes.

Cd behaves similarly to an essential element Zn, as they belong to the same group (column) in the periodic chart (Fig. 1.3). The chemical toxicity of Cd(+2) is well known (Chap. 11 in Ochiai (2008)), because it often replaces Zn(+2) due to its chemical similarity. By the way, Zn(+2) distributes widely in the body, as it constitutes many vital enzymes and DNA-controlling proteins. Hence, Cd, radioactive or not, would be distributed widely, like Zn. Cd(+2) also behaves like Ca(+2) in a way, as they carry the same electric charge and are of about the same size: 100 pm for Ca(+2) and 95 pm for Cd(+2). It is known that Cd(+2) concentrates in the bones.

Mercury (Hg) also belongs to the same group in the periodic chart (Fig. 1.3), and there are some similarities between Zn, Cd, and Hg, but Hg behaves significantly differently from Zn and Cd. There are a number of mechanisms in living organisms by which mercury compounds are eliminated from the body or made nontoxic (see, e.g., Chap. 11 in Ochiai (2008)). Radioactive mercury behaves likewise.

Selenium (Se) and tellurium (Te) belong to the same group as oxygen (O) and sulfur (S), the two essential elements. Se is quite toxic, but a few important enzymes and organic compounds contain Se as the active agent, and, hence, it is essential at a very low level. Tellurium is not known to be essential to any organism, and its behaviors in living systems are not well known. Radioactive Se and Te behave in living organisms in exactly the same manner as their nonradioactive counterparts.

Kr (krypton), Xe (xenon), and Rn (radon) are inert gases, and would not react with living matter in the chemical sense. However, they may physically stick to, say, mucous layers in the bronchi and lungs, where they can irradiate the surrounding tissues, thus, contributing to internal exposure. Radioactive Kr and Xe are β-emitters. Rn-222 is an α-emitter, which becomes (polonium) Po-214, which turns into (lead) Pb-214 through β-decay; both of these daughter isotopes can interact chemically with the cells and tissues. This point, i.e., that the daughter isotopes resulting in decay processes are different from the parent isotopes and can react chemically in addition to emitting radiation, is often neglected.

Uranium (U), radium (Ra), and plutonium (Pu) are metallic elements, and behave like other metals, such as iron (Fe) and molybdenum (Mo). Nuclear fuel consists of oxides of these elements. When uranium and plutonium are released into the atmosphere from bomb detonations or from the explosion of a nuclear power reactor, they would be in the form of U_2O_3, UO_2, UO_3, U_3O_8, PuO_2, etc., just like that of Fe, which is readily oxidized in air to FeO or Fe_2O_3. Uranium salts and plutonium salts in different oxidation states are mostly water soluble, and would behave like Fe, Mo, and similar elements. For example, Pu has been found to bind to an Fe-binding protein transferrin (Lehman et al. 1983), and Ra was found to bind to ferruginous proteins (ferritin and hemosiderin) (Nakamura et al. 2009).

(**NOTE**: all living organisms actually discriminate isotopes by their masses, but only to a very small extent. For example, plants absorb CO_2 for photosynthesis, but they slightly prefer $C^{12}O_2$ to $C^{13}O_2$ (both C-12 and C-13 are stable). This fact is utilized in order to make a judgment as to whether a certain ancient fossil that contains C was, indeed, of biological origin or not. There is also some discrimination between S^{32} versus S^{34}. However, this is not relevant to the issue of the health effects of radioisotopes.)

References

Fukuda T, Kino Y, Abe Y, Yamashiro H, Kuwahara Y, Nihei H, Sano Y, Irisawa A, Shimura T, Fukumoto M, Shinoda H, Obata Y, Saigusa S, Sekine T, Isogai E, Fukumoto M (2013) Distribution of artificial radionuclides in abandoned cattle in the evacuation zone of the Fukushima Daiichi nuclear power plant. PLoS One 8:e54321

Leggett RW, Williams LR, Melo DR, Lipsztein JL (2003) A physiologically based biokinetic model for cesium in the human body. Sci Total Environ 317:235–255

Lehmann M, Culig H, Taylor DM (1983) Identification of transferrin as the principal plutonium-binding protein in the blood serum and liver cytosol of rats: immunological and chromatographic studies. Int J Radiat Biol 44:65–74

Matsusaka N, Yamakawa Y, Sato I, Tsuda S, Kobayashi H, Nishimura Y (1997) Organ distribution of 137Cs in mouse fetuses and their dams. Radioisotopes 46:214–218

Nakamura E, Makishima A, Hagino K, Okabe K (2009) Accumulation of radium in ferruginous protein bodies formed in lung tissue: association of resulting radiation hotspots with malignant mesothelioma and other malignancies. Proc Jpn Acad Ser B Phys Biol Sci 85:229–239

Ochiai E (1977) Bioinorganic chemistry, an introduction. Allyn and Bacon, Boston

Ochiai E (2008) Bioinorganic chemistry: a survey. Elsevier, Amsterdam

Ochiai E (2011) Chemicals for life and living. Springer Verlag, Heidelberg

Chapter 11
The Effects of Radiation on Biological Systems

11.1 Expression of Dose Due to Exposure: Sievert

11.1.1 Radioactivity and Exposure Dose Expression Currently in Use

A source of radiation (radioactive material) emits radioactive particles of α, β, and γ, and the level of radioactivity is measured in terms of Bq, the number of decaying nuclei per second, which can usually be assumed to be equal to the number of radiation particles emitted. However, this is not exactly correct in certain cases. This value Bq represents, in a way, the number of radioactive nuclei present, as it is related to Bq = kN, where k is numerically a constant, indicating the probability per second of decay for a nucleus, and N is the number of that nuclei. The Bq value we will be dealing with will be something like the order of one to 10^{20}. If the radionuclide is Cs-137, $k = 7.3 \times 10^{-10}$ (see Table 2.2) and 10^{10} Bq represents 1.4×10^{19} Cs-137 nuclei, which is 2.3×10^{-5} mol = 3.2 mg. 10^{20} Bq represents 3.2×10^{7} g = 32 t (of Cs atoms). Similar calculations have been shown for other nuclides earlier (refer to Table 2.2).

Table 8.1 indicates that Cs released from the Fukushima Dai-ichi NPP was 3.3×10^{16} Bq (ignoring the difference between Cs-134 and 137), which corresponds to 10 kg of Cs atoms. The current Japanese regulations state that the allowable maximum radiation for food is 100 Bq per 1 kg. This means only 3.2×10^{-11} g of Cs (assuming that the only radioactive material in the food is Cs-137), which is a very tiny amount chemically, but significant in its radiation effects.

Radiation particles (including photons) emitted from a radioactive substance can impact on a material. If 1 kg of the material absorbed 1 J of energy from radiation, the dose (of the exposure) is defined as 1 Gy (Gray). This definition implies that 1 J of energy will be absorbed by the entire body of 1 kg of the material. The biological effect of radiation seems to depend on the character of the radiation; for example, an α-particle is heavy and electrically charged, and, hence, its ionization effect is

E. Ochiai, *Hiroshima to Fukushima*, Science Policy Reports,
DOI 10.1007/978-3-642-38727-2_11,

considered to be much stronger than that of β- and γ-particles of the same energy (Sect. 9.2). Hence, the biological dose is calculated by Gy times an effective factor Q, and this is called the "equivalent dose", which is expressed in terms of Sv (Sievert); i.e., $Sv = Q \times Gy$, and $Q = 20$ for α, $Q = 1$ for β and γ, and $Q = 10-15$ for neutrons. The reason for and scientific bases on how these Q numbers were determined are not clear, and are quite debatable, particularly in the case of internal exposure, as discussed later.

A simple calculation shows that 100 Sv on the human body would give 100 J/kg, if this radiation is due to γ-particles, and that it will raise the body temperature by 0.024 deg (assuming that the body's heat capacity is approximately 1 cal/g/deg). This small increase in temperature would not kill the person. However, radiation of 100 Sv will kill a person instantly. In other words, the effect of dose due to radiation exposure expressed in Gy or Sv would not adequately reflect the severity of the radiation.

Another problem is the definition; that is, the energy absorbed is defined as "per kg". This definition implies that the radiation energy absorbed spreads immediately over the entire body of, say, 1 kg. The argument above is based on this assumption. It is highly unlikely that the energy carried by radiation particles will spread evenly and so quickly in the form of heat (see, for example, Fig. 9.2). However, it is not possible to identify how a radiation particle will affect a body (by ionization, etc.) in detail. Therefore, this definition (Gy and Sv) may have to be used because there is no good alternative available, and it may be adequate for investigating the external and acute exposure (see Sect. 11.1.2 for a discussion on inadequacy, even in the case of external radiation).

When a radioactive material gets inside a body and then irradiates the surrounding tissue (internal exposure), as discussed in the next section, the reach of the radiation would be fairly local, particularly in the case of α- and β-radiation. Then, defining the radiation effect in terms of J/kg would be quite inadequate and misleading in regard to the radiation effect. For example, 2 mGy (or 2 mSv) means literally 0.002 J/kg. But suppose that it affects only about 10 g of the tissue where the radioactive material landed. Then, the effective dose would be 0.002 J/10 g = 0.002 J/0.01 kg = 0.2 J/kg = 0.2 Gy (=200 mGy ~ 200 mSv); that is, the effective dose (in the local site) is 100 times as high as the nominal value. It could be much higher or lower than this estimate; the true value cannot be determined easily.

11.1.2 Ambiguity in Exposure Dose Expression: Sv (Gy)

Sv (or Gy) is typically used for two different purposes: (a) the areal radiation rate for external exposure and (b) the internal exposure dose. Sv (Gy) values are computationally obtained from the radioactivity, Bq, from the source. The Sv (Gy) value of the areal radiation is defined as the energy deposited at a depth of about 1 cm in the human body. This value cannot be measured in the practical sense, and, therefore, the conversion factor from the measured value of the count per minute (cpm)) has been obtained in a model system or estimated based on several assumptions. In any case, this conversion factor is obviously dependent on the radionuclide. It has been estimated that 1 μSv/h = 108 cpm for Co-60 (γ) and 1 μSv/h = 120 cpm for Cs-137 (γ) (with a GM-10 instrument). This conversion factor is used to obtain Sv/h (Gy/h) from the measurements of cpm (or Bq). (Bq and cps (count per second) are not equivalent, because the efficiency of producing a count event needs to be taken into consideration, and, besides, a single nuclear decay (one Bq) may produce, e.g., 80 % β-particles and 20 % γ-rays.)

The other use for Sv is the internal exposure dose due to a one-time ingestion of contaminated material that has a certain Bq value. It is assumed (via the ICRP scheme) that this material, once ingested, will remain in a body according to the bodily scheme in which it is continuously discharged through urine, feces, and others (sweat, etc.). It is assumed that the discharge follows the exponential curve with a certain half-life (biological half-life). Meanwhile, the radiation activity (Bq) will decrease based on the physical half-life. Calculation by integration has been done theoretically on how much Sv will be deposited in the body through one's life due to the one-time ingestion of radioactive material. The conversion factor is then obtained to relate the Bq value of the ingested material to the Sv value, which is dependent on the nuclide and also the age of the person. In a certain case, sex is another factor. For example, the biological half-life of Cs-137 has been assumed to be about 70 days in adults, but, as seen earlier (Fig. 10.2), the discharge from organs does not closely follow an exponential curve. In other words, the conversion factors devised are not particularly adequate. Besides, in many instances, people would ingest contaminated material not just once but many times through drink and food, with different quantities each time, or even continuously. The Sv value in these situations can hardly be estimable, or would it be meaningful?

Other important dose (Gy/Sv) values have been obtained for people exposed to the atomic bombs in Hiroshima and Nagasaki (Sect. 11.3.4). These values have been estimated from a number of factors by researchers, and have been published and reevaluated several times, the most recent being DS02 (2009). How accurately these values for the atomic bomb survivors represent the real exposures may be difficult to evaluate, but it is quite doubtful that the real exposures are represented adequately. One way to reevaluate the doses for the survivors will be discussed later, in Sect. (13.1).

11.1.3 Exposure Effects, an Alternative Expression

Let's reconsider the issue that 100 Sv is the lethal dose, though it raises the body temperature by only 0.024°. Why is such a small amount of energy lethal?

Let's suppose that the radiation is mostly made up of γ-photons, each of which carries 1 MeV of energy, a typical value, and a body has absorbed the energy over

10 s. 100 Sv = 100 Gy over 10 s. 1 MeV is 1.6×10^{-13} J, and, hence, 100 J should be due to *ca.* 6×10^{14} photons/10 s. That is, 6×10^{14} γ-photons were absorbed by 1 kg of the body. There are about 5×10^{11} cells in 1 kg of the human body (see Sect. 3.3). This means that each cell received 1,200 γ-photons, on average. If a single γ-photon affects some 200 cells, then each cell will potentially be affected by 2.5×10^{5} photons. If the photons are more concentrated in a certain area rather than the entire 1 kg of the body, the effects on such local tissues and cells would be much greater. Then, there must be a great chance that many of the vital organs or tissues would be destroyed by these photons, and the repair mechanisms available could not cope with so much damage caused so quickly, and this would lead to the death of the person who received this much radiation. In terms of Gy = J/kg, the radiation effects are not obvious, but it seems to be clear if the radiation effects are expressed in terms of how many radioactive particles impinge on cells, tissues, or organs.

However, the health effects in the subsequent chapters will be discussed often in terms of the conventional Sv or Gy dose, just because it is the only available data.

11.2 External and Internal Exposure

The location was not specified in relation to the radioactive material when the dose due to an exposure to radiation was mentioned in the previous section. There are two situations here. One is that the radioactive material is outside of a living body, such that the radiation particles impinge on the surface of the body first, and then penetrate as far as they can. This is "external exposure". On the other hand, a radioactive material may get inside a living body, and then emit the radioactive particles within it. The radioactive particles will directly interact with tissues, organs, cells, and molecules inside the body. This is "internal exposure".

In external exposure, before arriving at the surface of a body, they (the radiation particles) have to travel some distance from the source, and will interact with molecules in the air and may even be stopped before reaching the skin. α-particles are likely stopped by the clothes covering a body, even if they manage to come close, and β-particles, even if they are able to penetrate the clothing, would not pass very deep into the skin. Hence, these radiations would have no significant internal effects, unless β-radiation has an enormously high energy. γ-radiation can travel over 100 m through the air, and, thus, arrive at the surface of a body, penetrate it, and, hence, is the only significant effecter in the case of external radiation. Another important radiation is the neutron, which can penetrate a body, as it carries no electric charge.

X-rays have been used for diagnostic as well as therapeutic purposes. This is a typical "external" radiation, as the source of the X-rays is obviously outside of the body to which they are applied. When the atomic bomb was exploded in midair over Hiroshima, it emitted an enormous amount of radiation. The immediate

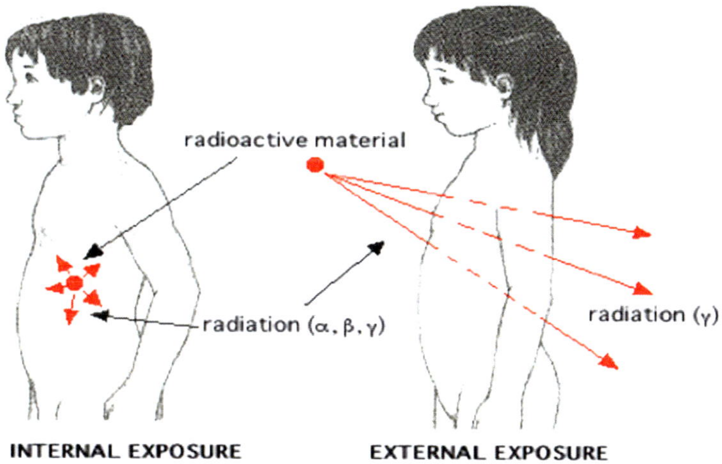

Fig. 11.1 Internal versus external exposure

radiation effect was "external", and this was considered to be the major radiation effect since the time of the Hiroshima bombing.

The difference between "external" and "internal" exposure is illustrated in Fig. 11.1. The atomic bombs killed many people (and animals) instantly or within a few months at the most. This is known an "acute" radiation effect. However, many more people have been affected by radiation, and they have suffered from not only cancers, but also various diseases and nondefinable illnesses. The exposure doses estimated by various means indicated that the exposure was rather low in the conventional sense, and, hence, many types of adverse health effect have not been recognized as radiation effects. It is now believed that the ill effects have been caused by the ingestion of radioactive material (of the fallout) through a number of routes: nose to lungs, mouth through esophagus to the digestive systems, or being absorbed into the circulatory system. In the cases of radiation related to the Chernobyl or Fukushima incidents, radioactive material could find their way into bodies through contaminated water/milk and foods, as well as fallout floating in the air. While the radioactive material is inside the body, it irradiates the cells, tissues, and organs; this is "internal" exposure.

The traditional radiation effects do not distinguish between "internal" and "external" exposure, and, as a result, do not represent the true severity of "internal" exposure. This comes from the traditional concerns about radiation, i.e., radiation caused by X-ray irradiation for the purpose of diagnosis and therapy, and the immediate (acute) effects of radiation originating directly from the exploded atomic bombs. These effects represent only "external" effects, though γ-radiation is assumed to affect different organs inside a body, with each organ being exposed to a certain assigned proportion of the entire dose (ICRP).

11.3 Radioactive Material, Natural and Artificial, and Background Exposure

11.3.1 Radiation Sources

A large number of radioactive isotopes are known. The most commonly encountered radioisotopes and radiation sources are as follows. The first two (a and b) are natural sources and the last two (c and d) are artificial sources.

(a) Naturally occurring radioactive isotopes (half-life, radiation particle emitted) on the current earth include: U-234 (half-life = 2.5 × 10^5 years, radiation α), U-235 (7 × 10^8 years, α), U-238 (4.5 × 10^9 years, α), Th-232 (1.4 × 10^{10} years, α), Ra-226 (1,600 years, α), Rn-222 (3.82 days, α), K-40 (1.3 × 10^9 years, β), C-14 (5,730 years, β). U, Th, and K-40 have very long half-lives, and, hence, they are still present on this earth today. Most of the radioisotopes present upon the formation of the earth (4.6 billion years ago) have decayed and disappeared. Ra, Rn, and Po (as well as some radioactive Pb) are decay products from U-238, and, hence, are still present today. C-14 has a short half-life and is not what remains from the beginning of the earth. Rather, it is constantly created from N-14 (a stable isotope present on the earth) by the bombardment of neutrons from the Sun (i.e., cosmic rays), $_7N^{14} + {_0n^1} \rightarrow {_6C^{14}} + {_1p^1}$, where $_0n^1$ is a neutron and $_1p^1$ is a proton. A tiny amount of $_1H^3$ (tritium) is also being produced in a similar way.

(b) Radiation falls on the earth from the universe. The universe is full of radiation (cosmic rays) of various kinds, and some of them reach the earth's surface. The most important ones are various radiations from the Sun. Obviously, visible light and some parts of ultraviolet light are actually beneficial to the organisms on the earth. Rather, it is indispensable. The Sun emits all the other kinds of radiation as well: α, β, γ, and neutrons. The electrically charged particles (α and β) do not often reach the surface of the earth, because their paths would be curved away by the magnetic field around the earth. Neutrons and photons (γ) should proceed in a straight line, but these interact with the molecules in the atmosphere and are largely stopped. That is, not much reaches the surface of the earth.

However, they say that you would receive more cosmic rays when you are flying. An estimate indicates that being 10 km above the surface of the earth may expose you to about 3.9 μSv/h. So, if you flew naked at that height for 10 h, you will receive a dose of 39 μSv. But if you were in an airplane, the dose would be much smaller.

(c) Radioactive isotopes that form in the nuclear fission reaction (of U) include: Pu-239 (2.4 × 10^4 years, α), I-131 (8 days, β), Cs-134 (2.06 years, β), Cs-137 (30 years, β), Sr-90 (28.8 years, β), Tc-100 (16 s, β), Zr-93 (1.5 × 10^6 years, β); see Table 2.1 for more data. Altogether, about 200 radioisotopes are known to form in the nuclear fission reaction. But in many instances, the majority of radioactive material spewed out of a reactor are radioactive isotopes of inert gases such as Kr-85 (β) and Xe-133 (β). These are not reactive (chemically), but can easily enter a body through inhalation, and can have serious effects, but they are often ignored.

(d) In addition, industrial, diagnostic, as well as therapeutic applications of radiation are utilized in today's world. They are mostly X-rays and γ-rays.

11.3.2 Background Radiation

As seen in the previous section, a number of naturally occurring radioisotopes are present in the environment. They are not equally and universally distributed on the present earth, but all living organisms are more or less subject to radiation from these sources. In other words, all organisms are exposed to the "background" radiation. Background radiation includes: (a) radiation from natural sources, mostly external exposure to their γ-rays, (b) internal radiation from ingested naturally occurring radioactive material, and (c) cosmic rays. Natural radiation (a) varies from one place to another, depending on the distribution of radioactive U, Th, Ra, etc. (for example, Fig. 11.2 shows the uranium distribution in the US). There are two different kinds of internal radioactive material. K-40 and C-14 are universally present, more or less equally distributed in the environment, and are also present at constant levels in living organisms. Minute amounts of naturally occurring U, Th, Ra, etc. may be present in human bodies, but they vary in quantity. The exposure to cosmic rays also varies. In other words, the exposure to natural radiation is not constant, i.e., it is not the same for everybody.

The internal exposure due to K-40 and C-14 can be easily estimated, as their natural abundances are known and the contents of C and K are well established. First, the content of K in the human body is 2,200 ppm; hence, the amount of K in the human body is 2.2 g/kg body. K-40 abundance is 0.012 %, and, so, K-40 is present at 2.6×10^{-4} g/kg. The half-life of K-40 is 1.3×10^9 years, and, hence, k (rate constant) is $\ln 2/(1.3 \times 10^9 \text{ y}) = 1.7 \times 10^{-17} \text{ s}^{-1}$. The number of K-40 is $(2.6 \times 10^{-4}/40) \times (6.02 \times 10^{23}) = 3.9 \times 10^{18}$. Therefore, it would give rise to 66 Bq/kg (β and γ mixture). The energy of the decay of K-40 is about 1.3 MeV and 66 Bq of radiation keeps impinging on 1 kg of body tissue, so it would amount to $66 \times 60 \times 60 \times 24 \times 365 = 2.08 \times 10^9$ β-/γ-particles of 1.3 MeV per year, assuming that all the energy is absorbed by the 1 kg of the body. This is equivalent to 0.43 mJ/kg/year or 0.43 mSv (/kg)/year. Some portion of the radiation may leak out of the body, and, hence, the dose may actually be somewhat lower.

The human body contains about 190 g of carbon per kg. About 2×10^{-10} g is C-14. A calculation similar to that above gives 0.03 mSv/year. The total dose due to the internal exposure to K-40 and C-14 is 0.45 mSv/year. This is almost constant and everybody everywhere is exposed to this much internally from these sources. Other sources vary, and one estimate by a UK authority is given in Table 11.1 (p. 64

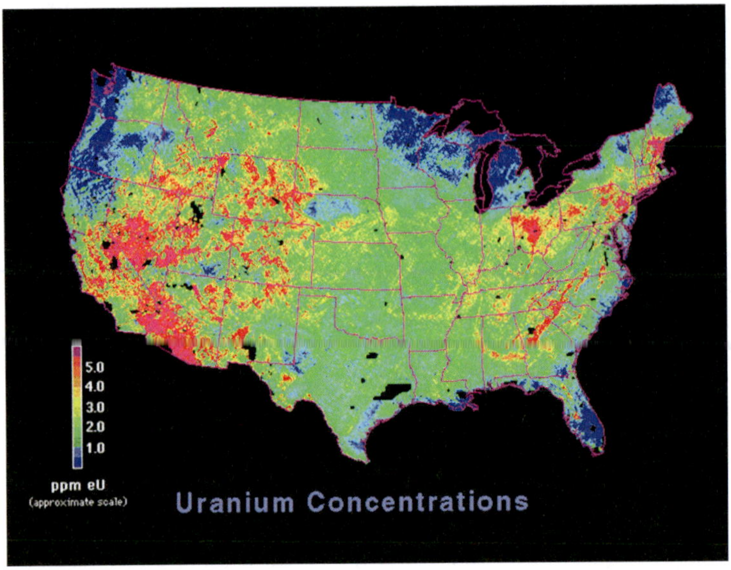

Fig. 11.2 Uranium concentration in the United States of America

of ECCR 2010). The internal exposure dose due to K-40 in this estimate is somewhat smaller than the value obtained above. This might be caused by using a different method of converting Bq to Sv. The values given in the table are considered to be about the average background radiation to which a human being is subjected around the world. The average annual dose in Japan was considered to be lower than this, and was about 1.0 mSv/year before the Fukushima Dai-ichi NPP accident. A problem with this kind of estimation is that internal and external exposure doses are not distinguished, and, hence, they are given an equal significance. This is not the case, as will be discussed shortly.

Human beings have survived this much radiation throughout mankind's history. However, it could be that the background radiation has significantly increased during the last century or so because of humans' activities with radioactive material, and, so, the background radiation level might have been somewhat or significantly lower in the past than the current level (see Sect. 16.5).

No matter what the case might be, it implies, perhaps, that the human body has several mechanisms to repair the damage caused by this much radiation (i.e., background). It has been suggested, though, that even this much background radiation exposure alone may be causing about 1 % of cancers if humans are exposed to it over a period of 70 years (Caldicott 2011). Recently, UK researchers (Kendall et al. 2012) concluded, after careful studies of 27,000 cases in the period 1980–2006, that about 15 % of the 500 or so cases of childhood leukemia that occurs annually in the UK is due to the natural background γ-radiation. They

Table 11.1 Annual effective background dose (UK population)

Source	Average (mSv/year)	Range (mSv/year)
Internal K-40	0.165	
Internal C-14	0.012	
Internal U and Th	0.12	Variable
Radium, radon, and daughters	1.1	0.3–100
Thorium daughters	0.09	0.05–0.5
Cosmic ray secondaries	0.28	0.2–0.3
Cosmic ray neutrons	0.1	0.05–0.15
External terrestrial	0.48	0.1–1.0
Total	2.35	1.0–100

estimated that the extra risk rate (ERR) is 12 % (range 3–22 %) per mSv in childhood leukemia due to background radiation on bone marrow.

By the way, according to the IAEA, 1 kg of soil typically contains 370 Bq of K-40 (typical range 100–700 Bq), 25 Bq of Ra-226 (10–50 Bq), 25 Bq of U-238 (10–50 Bq), and 25 Bq of Th-232 (7–50 Bq).

11.3.3 Artificial Sources of Radiation: Medical Uses

Table 11.1 lists the radiation exposures due to natural sources, but normal human beings in today's world are exposed to a number of sources of artificial radiation, as listed above. First, in today's medicine, X-rays are quite often used mostly for diagnostic purposes. Particularly, computed tomography (CT) scanning irradiates a body for long time and over a wide range, and emits as much as 7 mGy (mSv) per scan. A regular diagnostic X-ray gives a dose of about 0.05–0.5 mGy (mSv). X-rays can also be used for therapeutic purposes. Essentially, it is targeted at the cancerous tissue, and destroys (kills) the cancerous cells.

γ-Radioactive material can be used for diagnostic purposes, as a tracer. For example, a solution containing Tc-99m is injected into a vein. When it circulates the body, γ-ray monitoring (scanning) will locate abnormalities in the blood flow. I-131 was also used for this purpose, although it can get into the thyroid gland as well. One type of therapy implants a source of radiation in a cancerous tissue, which aims to destroy the cancerous cells. This is a typical "internal" irradiation.

The health effects of radiation were first revealed regarding the use of X-rays as a diagnostic tool. In 1955, Dr. Alice Stewart at Oxford University became aware of a sharp rise in leukemia among young children in England. By 1957, she had completed a survey of about 1,300 mothers, half of whose children had leukemia and other cancers. It turned out that babies born of mothers who had undergone a series of X-rays in the pelvic region during pregnancy were twice as likely to develop leukemia or other cancers as compared to those born of mothers who had received no X-rays. The rate of developing leukemia was dependent on the number of X-ray tests received (Stewart et al. 1958). This implies that fairly low doses of radiation can affect the fetus, which may later develop leukemia or another cancer, and that the effect is cumulative. There are now a large number of reports that demonstrate the increased risk of cancer linked to the diagnostic use of X-rays.

11.3.4 Atomic Bombs

The problematic source is radiation from nuclear weapons and nuclear power reactors. Let us first look at the data on the Hiroshima/Nagasaki atomic bombs. Refer to Chap. 5 for the extent of the energy produced and the physical devastation caused by the atomic bombs in Hiroshima and Nagasaki. Scientists at the Radiation Effects Research Foundation (RERF, Hiroshima) and its predecessor, the Atomic Bomb Casualty Commission (ABCC), have attempted to estimate as accurately as possible the dose due to the γ and neutron radiation arising from the atomic bomb explosions, based on a number of data, measurements of remaining artifacts, experimental measurements, etc. The atomic bombs certainly produced a large number of radioactive nuclides, as well as neutrons. Of these, α and β radiations were discounted, because they would be quickly deactivated and stopped via collisions with molecules in the air and, even if they reached a human body, they would not penetrate it significantly. The latest data available were published in 2006 as DS02 (Young et al. 2003). Its main findings are illustrated in Fig. 11.3, drawn from the published data. The dose at the hypocenter is estimated as being 350 Gy in Nagasaki and 165 Gy in Hiroshima. At 1 km from the hypocenter, the dose in Nagasaki was 8.7 Gy and 4.5 Gy in Hiroshima. In the case of Hiroshima, a strong typhoon (known as the "Makurazaki" typhoon) arrived in the area on September 17th, 1945, a month after the atomic bomb, and is believed to have washed away a significant portion of the radioactive material into the surrounding sea. This may have made the radiation dose of Hiroshima appear lower than the real values.

Initially, the US authorities ignored the effects other than those mentioned above. However, it has become apparent that the fallout of the atomic explosion did, indeed, have radiation effects on the survivors. It is highly likely that the fallout (radioactive material) could have had external effects, and, also, somehow, gotten into and internally irradiated their bodies. It has had long-lasting and slow-emerging health effects on many survivors. This issue will be discussed shortly.

This issue is also related to the fallout created by the nuclear explosion tests conducted by nuclear nations: US, USSR (now Russia), UK, France, China, India, and Pakistan. At least 2,000 above-ground tests have been conducted (Sect. 4.3, Fig. 4.4). These tests have spread all kinds of radioactive nuclides over the world and affected human health significantly, though this fact is not well known (see Sect. 16.1).

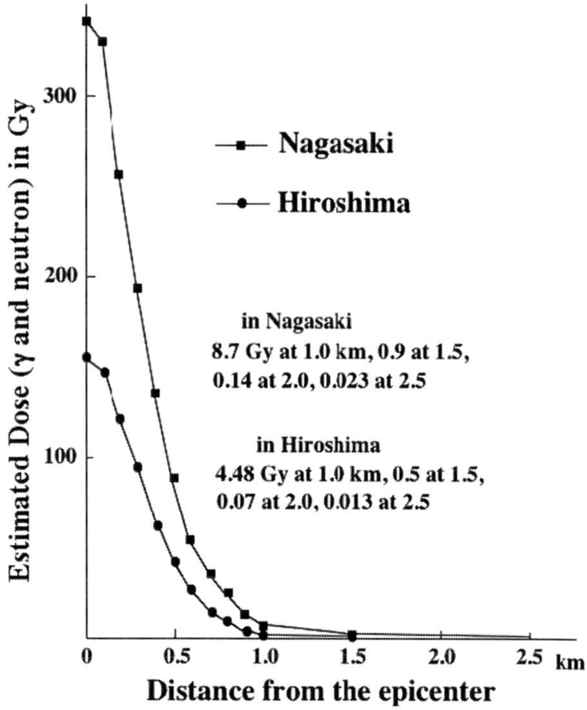

Fig. 11.3 Radiation dose in Hiroshima and Nagasaki (Data are from "Hiroshima survivor dosimetry 2002 DS02", RERF 2006)

11.3.5 Radiation Escaped from Nuclear Power Reactors

The currently most serious issue is radiation released from nuclear power plants and others, which contains nuclear fission products (of U-235 and/or Pu-239), such as Cs-137, Sr-90, and I-131, as well as U-235/238 themselves, and Pu-239 (and other transuranium nuclides). The serious problems are "internal" radiation due to these materials that have somehow entered the body. Most of the important nuclides concerned are listed in Table 2.1, along with their characteristics. The health effects of these radioisotopes will be discussed shortly.

11.4 Acute Radiation Effects

11.4.1 General Trends

One of the fundamental trends in radiation effects on living organisms is that the tissues/cells which are rapidly proliferating, through both mitosis and meiosis, are the most vulnerable to radiation effects. It is believed that when cells, particularly the starting stem cells, are damaged in the process of proliferation, lesions may accumulate as they divide and, hence, the lesion is multiplied. Therefore, the cells which are the most sensitive to radiation include lymphocytes, erythroblasts, intestinal cells, and spermatogonia (the stem cell for sperm production). Fathers who have mutated DNA in their sperm may transmit the mutations to their offspring. Another active tissue is hair follicles, which are quite radiosensitive. So, a moderate dose can cause hair loss, which could be temporary or permanent, depending on the dose. Skeletal muscle cells, nerve cells, and cartilage cells are much less sensitive.

In terms of age, the young are more sensitive to radiation, and, thus, vulnerability is at its highest in the embryo/fetus (the earlier the more sensitive), followed by newborns, babies, and children. Hence, pregnant women and babies/children must be very carefully protected from radiation.

The acute effects of radiation on animals are summarized in Fig. 11.4. This is entirely an "external" radiation effect and based on the experimental data (dogs, etc.). It shows the average survival time when the subjects receive one shot of high external dose, though the sensitivity somewhat varies among species.

This figure represents an average. The main cause of death seems to depend on the dose. The hematologic (blood-related) causes are the most major below 10 Gy, and the exposed subjects survive, on average, for several days to several weeks. Over the range of 10–50 Sv, most subjects died of failure in the digestive tract within several days. At higher doses, brain damage seems to be the major cause, and the exposed suffer from instant death or death within a few days. This represents only the typical and likely effects of high external doses, and should not be considered to be exclusive of other causes. Besides, internal exposure effects due to ingested radioactive material would be quite different, though the vulnerability of different organs and tissues is represented by this figure.

It has been shown (Chap. 4 in Forshier 2009) that the initial reaction to external radiation is anorexia, nausea, vomiting, and/or diarrhea, even when the dose is lower than 2 Gy (2 Sv). It lasts for a few minutes to a few days. This is then followed by a latent period; though changes are taking place, they do not yet become apparent. Then, noticeable effects become manifest, and the affected body would die or recover in some cases. The length of time taken for these processes varies depending on the dose.

The most vulnerable seem to be bone marrow and blood-producing tissues. At first, no significant drop in blood cells occurs, though the bone marrow stem cells are dying, because blood cells last 120 days or so. After a latent period (3–5 weeks), the blood cell count starts to decrease. If the exposure was 1–3 Gy or less, the bone marrow could replenish enough blood cells to sustain life in many individuals, who usually recover fully within 6 months. Above 3 Gy, the survival rate decreases. Leukemia can also appear at lower doses, and will be discussed later.

The next vulnerable part of the human body is the lining cells of the gastrointestinal tract, which are also constantly reproducing themselves. Radiation in this area first causes nausea, vomiting, and diarrhea, and then destroys the villi of the lining cells. This lessens the absorptive capacity of intestinal cells. Death arises from

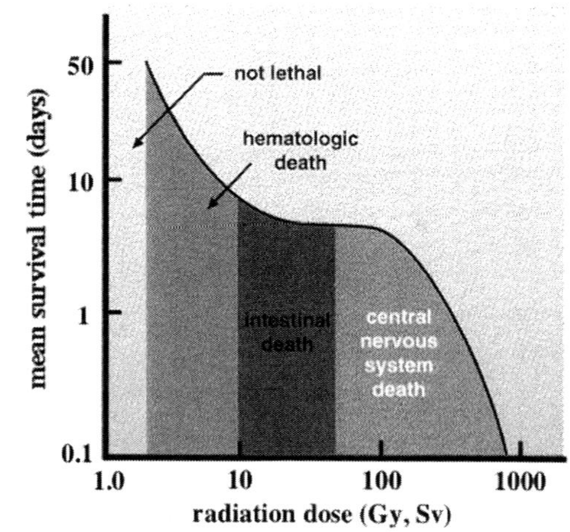

Fig. 11.4 Survival rate dependence on the whole-body external dose (Adapted from Forshier 2009)

dehydration and infections in the area. Brain death seems to be caused by elevated fluid pressure due to circulating systems in the brain; however no details are known.

The skin is obviously the most vulnerable to external radiation. It is composed of three layers; the surface epidermis, the middle layer connective tissue dermis, and the subcutaneous layer. The epidermis is made of the outermost mature cells that are nondividing and a portion of the immature dividing cells (epidermal basal layers). Skin keeps shedding the outermost portion, about 2 % per day, and the basal layer provides replacement. Hence, this dividing portion is quite radiosensitive. Hair follicles are present in this layer, and, hence, are easily destroyed, resulting in hair loss.

Male reproductive organs by themselves are not very sensitive to radiation. But one particular kind of cells in the testes, spermatogonia, are rapidly dividing, and, hence, are very sensitive to radiation. This cell turns into sperm through the stages of spermatocyte and then spermatid. Experimental studies indicated that a dose of 2–2.5 Gy makes testes sterile for about a year before they recover. Higher doses cause permanent sterility. Some mutations in the spermatogonia caused by radiation could be transmitted to the next generation. The female reproductive system is less sensitive to radiation, but the ovum (egg) may be subject to radiation, resulting in mutations of DNA, which may manifest in the born child.

Cataracts are due to damage to the lens in the eye. This may be caused by radiation, even ultraviolet light. Cataracts are, thus, common among elderly people not necessarily exposed to radiation, but is a prominent phenomenon in the radiation-exposed population.

11.4.2 Atomic Bomb Victims

The atomic bomb explosion on the city of Hiroshima is estimated to have released 6.3×10^{13} J of energy, of which 15 % was in the form of radiation (see Chap. 5). As shown in the previous section (Fig. 11.3), the dose (to humans) was estimated to be about 170 Gy at ground zero and 4.5 Gy 1 km away in Hiroshima, and 350 Gy at ground zero and 9 Gy 1 km away in Nagasaki. People who survived the physical effects (i.e., the heat and windblast) of the explosion died from exposure to these high doses (perhaps higher than *ca.* 10 Sv or so) of radiation, either instantly or within a few weeks, likely due to internal organ failure, as discussed in the previous section. It has been estimated that 60–160 thousand people died within 2–4 months; how many of these died from acute radiation effects only is not known. However, these acute effects are obvious, visible, and tangible, and cannot be denied by the authorities. Why such a small amount of energy (100 Gy = 100 J/kg) is fatal was previously discussed in terms of the interactions of radiation particles with cells, tissues, and organs (Sect. 11.1.2).

It has been estimated by the authorities that exposure to radiation of less than 1 Sv may not have been fatal, but caused nausea, vomiting, hair loss, and some loss of white blood cells. Doses less than 0.25 Sv are supposed to have caused no significant immediate harmful effects, but they did cause slow-manifesting illnesses, including various cancers. The slow-developing effects will be discussed in later sections.

To illustrate the immediate acute radiation effects, an eyewitness account is quoted below. It is taken from a talk given at the opening press conference of the "Association for Citizens and Scientists Concerned about Internal Radiation Exposures" (ACSIR) on January 27, 2012. The speaker is Dr. Shuntaro Hida, who is himself an atomic bomb survivor and witnessed the horror of the effects of the atomic bomb as a physician.

> In 1945 at the age of 28, I was serving as a medical officer in a Hiroshima military hospital. On August 6, I happened to be staying at the hospital, but was called away unexpectedly around 2:30 am to examine a farmer's child who had suffered a fit caused by valvular heart disease. .... As a result, on the morning they dropped the atomic bomb, I found myself in Hesaka, a village six kilometer northeast from the epicenter. .... That afternoon while I was still in Hesaka, a steady stream of surviving hibakusha (explosion-affected persons) arrived, seeking refuge in the tiny village. ....There was simply no place to shelter them. In the end, the seriously wounded who had fled from Hiroshima, burnt from head to toe, with burnt and festering skin, had no alternative but to lay themselves down on the road, on the school ground, and in empty land plots in the village.
>
> ... That evening by chance, four more medical officers happened to be present in the village. ...6000 victims spent that night in Hesaka. ... However, as there was no medicine and no medical instruments, and we medical officers were forced to simply stand by as more and more of the wounded died. ....
>
> Until the fourth morning, we had assumed that all those that died had succumbed to their burns. However, beginning that morning, patients began to die who had not suffered serious burns. ... By that morning, 27000 victims had entered the village. ...
>
> ... But what sort of symptoms were they afflicted with? First a high fever of 40 degrees. A fever so high that even doctors of internal medicine had rarely seen it. ... as we examined

our patients and wondered why they were running such fever, they began to bleed from their eyes, nose and mouth. Even we doctors had never seen such bleeding from the eyes.... we attempted to examine the inside of their mouths, but could not. It was not simply bad breath, it was the smell of decay. A smell so bad, we could not put our faces near their mouths....even though these people were still alive, the insides of their mouths were decaying.

...In time, purple spots began to appear on the undamaged portions of our patients' skin....Finally hair loss. ...They would gently touch their heads, and the fairs would fall away....After these rare symptoms appeared, the patient had at most an hour or two to live....This cluster of symptoms would later be termed 'acute radiation syndrome'....

11.4.3 Other Incidents

An unfortunate incident where acute radiation exposure killed workers happened at a nuclear fuel reprocessing factory of a company called JCO in Tokai village, Ibaragi prefecture, south of Fukushima on September 30, 1999 (Tokaimura nuclear accident, Wikipedia). The workers did not follow closely the appropriate instructions and cut corners in their treatment of concentrated uranyl nitrate solution. It caused the uranium concentration to become slightly supercritical because the solvent water played the role of a neutron moderator but the concentration was too high. Apparently, the criticality continued on and off for about 20 h. It released neutrons and other radiation. The two workers closest to the container of uranium immediately experienced pain, nausea, and breathing difficulty, etc. One who received a dose of 16–20 Gy died from the failure of multiple organs 83 days later. Another who received a dose of 6–10 Gy also died of organs failure 211 days later. Another who received a dose of 1–4.5 Gy lost most of his white blood cells, but recovered following bone marrow transplant. Similar incidents have happened at various installations in the US, as well as others.

Some of the people who worked at the damaged Chernobyl reactor (unit 4) immediately after the accident were exposed to strong radiation, as the accident was a sort of nuclear explosion and spread an enormous amount of radioactive material, not only fission products but also uranium itself and transuranium nuclides such as Pu-239. They had shown acute radiation symptoms, and some died within a few weeks. No accurate official data on the deaths among the cleanup workers (liquidators) were published, as the data were suppressed by the Soviet Union authorities. It has been found among civic records, which escaped official confiscation, that even civilians living near the reactor showed acute symptoms (Lupandin 1998). The exposure dose among people in the surrounding area has been estimated to be rather high, with more than 15 % of them receiving more than 1 Sv. This estimate did not include internal exposure.

11.4.4 Acute Internal Exposure Effects: Atomic Bombs

A very small quantity of radioactive material can enter the body via many routes. Minute particles (as fallout) containing radioisotopes or gaseous entities such as I_2, Kr, and Xe may be floating in the air after an atomic bomb explosion or an accident like that of Chernobyl or Fukushima. A small quantity of gaseous material or minute radioactive particles may be breathed in through the nose or mouth, and then enter the bronchi and then the lungs, and may be settled there. Radioisotopes may get into water or soil, and then into plants, which may be eaten by cattle or others. Or they may get into planktons, sea plants, and fish. Eventually, these can get into human bodies through drinking water, milk, or food.

Let us hear again what Dr. S. Hida had to say about the likely internal exposure effects seen among the immediate atomic bomb survivors. This is a quote from the speech mentioned in Sect. 11.4.1.

> What next surprised us was the patient who claimed, "Doctor, I'm not sick from the 'pika' (the atomic bomb explosion)." "What makes you say that?" I queried, and he answered "Well, I wasn't in Hiroshima that day!" He continued "I didn't come to Hiroshima until two days after the August 6th bombing. You see, one of my children did not return home after the bombing, so I traveled to the city to look for him. It wasn't until after walking around the ruins for two days that I began to feel ill, and that's when I thought I should come here and get checked out by you.
>
> With all our medical knowledge combined, we could not figure out what was wrong with this man. . .... Soon after, he began to display a number of odd symptoms and passed away. "What on earth killed that man?" we wondered. "If it wasn't the bomb. . .. Then what? This turned out to be my first encounter with radiation sickness.
>
> Then, we saw what we now call "internal radiation exposure". . . ..Back then, we called it "city-entrance exposure". These individuals were not in Hiroshima at the time of bombing; they had come in the city later. . ..and then fell ill. . ...For the life of us, we could not figure out why. . .. . .

So, these people who had not been exposed directly to the radiation from the bomb itself but had entered the city days later, nonetheless, exhibited symptoms similar to the direct exposure and many died soon after. They must have somehow ingested significant amounts of radioactive material still scattered and floating around in the city, and then suffered from the acute effects of internal radiation exposure.

11.5 Low-Dose Internal Exposure Effects

External exposure at low doses may not affect health significantly; as mentioned earlier, lower than 0.25 Sv (250 mSv) is said to have no immediate effects. Even today, the Japanese government insists that such a low dose as 20 mSv/year would have no ill effects. Also, the ICRP suggests that there is no cancer-causing effect below 100 mSv. And so on. In these claims, there is usually no clear distinction made between external and internal exposure. What they have in mind is "external" exposure, as their recommendation is based on the data related to the atomic bombs. These arguments may be true if, indeed, one is exposed strictly only "externally". However, to extend these numbers to "internal" exposure would not represent the true nature of radiation effects. In fact, many atomic bomb survivors who were not killed at the time of the bombing and were seemingly healthy have, however, suffered from various adverse health effects for a long time. We will discuss these issues here and present the relevant data in a later chapter, but first, we gather some ideas about how and in what form the fallout from atomic bombs as well as that released (accidentally or intentionally) from nuclear power plants would behave.

11.5.1 How and in What Form the Radiation Particles Were Spread from the Explosion of an Atomic Bomb?

In Hiroshima, the atomic bomb exploded at a height of 600 m above the ground. Because of the enormous heat created, the debris from the explosion had mostly risen up before falling, but, simultaneously, a very strong wind was caused, which spread the fallout in all directions. While the fallout was in the air, it was emitting a huge amount of radiation. Some people may have ingested such floating particles (causing internal effects), but the radiation effects on most of the people at this stage might have been external exposure to neutrons and γ-rays. The distribution of dose (due to external exposure) as a function of distance from the epicenter is shown in Fig. 11.3. In the case of external exposure, the chemical forms of radioactive material would not matter, but the chemical forms are critical when they get into the body.

The contents of the Hiroshima atomic bomb was essentially U-235 (plus U-238), in the form of oxide (UO_2). At the instant of explosion, most of the chemical material vaporized because of the high temperature. But as soon as the material was thrown upwards, it was cooled and formed condensed material, mostly in the form of minute particles (aerosols), though some of the fission products remained as gas. Such entities include iodine (I_2) and inert gases such as Kr and Xe, and are floating as independent molecules (atoms).

Other elements are likely converted into oxides as they encounter oxygen in the air; Cs_2O, SrO, MoO_2, TcO_2, PdO, Ag_2O, CdO, PuO_2, UO_2, etc. Some of them might have combined with CO_2 in the air to form carbonate, e.g., $SrCO_3$, or embedded in various aerosols made of sulfate and others. However, what forms they take and how they behave have not been well studied; it is rather difficult to study these features. Most of them would likely form minute particles, maybe on the order of micrometers or even smaller, and, hence, be readily carried by wind. Larger and denser particles may tend to come down sooner. Even minute particles eventually attach themselves to larger particles (not necessarily from the fallout) and would fall down. The range of spread could be fairly large; the fallout could spread over the entire surface of the earth, depending on the wind direction and other atmospheric conditions.

As mentioned earlier (Chap. 5), soon after the explosion of the atomic bomb, the so-called "black rain" fell on a wide area of Hiroshima, as well as Nagasaki. Water vapor must have condensed on the minute particles, forming raindrops. It is very likely that the adjective "black" of the rain implied that it contained a lot of fallout debris, washed down significant portions of radioactive particles, and contaminated the rained area. It was widely reported that, afterwards, many people saw a large number of dead fish floating on the rivers and ponds in the rained-on area. Once dried, the minute particles would be blown out of the ground and float again. It must have affected humans as well, either directly or via internal exposure. The internal exposure at relatively high doses would have manifested as acute syndrome, described above as "city-entrance exposure". The radiation dose mentioned earlier (see Fig. 11.3) does not take account of the internal exposure due to the fallout.

11.5.2 Release and Spread of Radioactive Material from a Nuclear Power Plant Due to an Accident

The behaviors of radioactive materials released from a nuclear reactor due to an accident are dependent on the nature of the accident. The worst case is a nuclear explosion. This happens when a critical condition is somehow brought about. The explosion may not be as severe as the atomic bomb explosion, but it would exert a strong physical power to destroy the building housing the reactor, as happened at the Chernobyl NPP and unit 3 of the Fukushima NPP. Depending on how far the explosive power extends, it can release most of the material in the nuclear fuel rods, along with neutrons and water vapor. The presence of water would change the chemical nature of the released isotopes; oxides may turn into hydroxides, e.g., Cs (OH), $Sr(OH)_2$, etc. But the overall nature of the aerosols may be the same as that found in the fallout of an atomic explosion.

Hydrogen explosions and steam explosions would spread mostly gaseous products such as Kr, Xe, and I_2, along with some water-dissolved fission products, such as Cs(OH). Water-insoluble material would also be splashed out, but not as much as water-dissolved species. The gaseous pressure of the reactor was relieved by venting it; this likely happened at units 1 and 2 at Fukushima, which released a high level of Kr, Xe, and I_2.

The form of radioactive material, mainly Cs-134/137, released from the Fukushima Dai-ichi NPP was studied by Kaneyasu et al. (2012). They assumed that it was either Cs(OH) or CsI when released, but it was then embedded into some form of aerosol. They collected particles first on 2012.04.28–05.12, and then on 2012.05.12–05.26. The amount collected was minute; a few femtograms in a sample of 1 m^3. The distribution of particle sizes was found to have maxima at 0.2–0.3 and 0.5–0.7 μm, with the majority around the second maximum. The floating soil particles or sea salt particles are known to be of sizes larger than 2 μm. The size distribution of sulfate aerosol seems to fit that of the Cs particles. Besides, it has been shown that the size of sulfate aerosols has two peaks, with the larger one forming as a result of reactions of the smaller particles with carbon dioxide. Hence, they suggest that the cesium from the Fukushima NPP has been spread as sulfate aerosol of size 0.2–0.7 μm or so.

The particle size of the suspended aerosol in the Chernobyl area seems to be one order of magnitude larger than the particles from Fukushima (Garger et al. 1998). The samples of aerosol were collected during the period from 1986.09 through 1993.06. The maximum in the size distribution occurs at about 4 μm in the range 2–30 μm. The mechanism of this aerosol formation consists of two processes: local resuspension (of deposited aerosol) and advective transport of the aerosol from highly contaminated areas. During a forest fire at a distance of 17 km, the majority of the radioactivity was associated with smaller particles in the range 0.28–0.50 μm. The size difference between the radioactive aerosol (of Cs) of Fukushima and Chernobyl could be due mostly to the climate differences between the two locations.

As the Fukushima Dai-ichi NNP is located on an ocean coast, the circulating contaminated water leaked out into the ocean, due to the malfunctioning water circulating system and, also likely, some leakage of the reactor due to the earthquake. Hence, the seawater in the immediate vicinity of the plant was highly contaminated, and the radiation level there is still high about 2 years later (at the time of writing), though it would have spread thinly through the mixing effects of waves and currents. It seems that some isotopes such as Cs and Sr are concentrated in the soil at the bottom of sea. Seaweed and fish in that region are still contaminated.

11.5.3 How Radioactive Material (Gas or Small Particles) May Enter the Human Body and How They Behave?

Radioactive material comes in various forms, as discussed above. The gaseous material I_2 (I-131), Kr-85, Xe-135, and Rn-222 (but Rn is not a fission product) can be inhaled. It will go through the nose, trachea, bronchi, and alveoli, and then into capillaries (blood vessels) in the lung, much like the path taken by air. Minute particles (aerosols) of other radioactive material may also enter similarly, but they would have a greater tendency to be stuck in the mucous layers somewhere along the way. Eventually, they may pass through the membranes of alveoli, then enter the capillaries, and may then circulate throughout the body and be somewhat settled somewhere. Kr, Xe, and Rn would remain as a gas dissolved in body fluid and fat, but their decay products could be solid in the forms of Rb^+, Cs^+, Fr^+ salts, and may persist. I_2 very likely turns into I^- (negative ions) when it comes into contact with water (through $I_2 + H_2O \rightarrow HI + HIO$; HIO may react further to turn into I^-).

How about aerosols containing Cs-134/137, Sr-89/90, etc.? It could get directly into a body through inhalation (via the mouth or nose), or first settle in soil and then be transferred into plants. Or it may get into water systems, directly or indirectly through soil. From there, contaminated plants would be eaten by cattle, and be transferred into milk and meat, i.e., contaminate them. Then, a person eats the contaminated plants, meat, and fish, and drinks the contaminated milk and water. This is just a conceptual description. Where, how much, in what way (through plants, etc.), and when, etc. cannot be specified. They can only be determined by the systematic and continuous monitoring of all kinds of material: air, soil, water, food, etc.

11.5.4 Basic Mechanisms of Internal Exposure Effects

The basic mechanism is the interaction of radioactive particles (including γ-photons) of high energy with molecules present in the cells, tissues/organs, and body fluid. The modes in which interactions occur are discussed in Chap. 9. Essentially, they eject electrons, leaving the molecule in an electron-short condition, i.e., ionic or free radical. Or it excites a molecule, of which a bond may be cleaved, leading to decomposition. The radiation particle loses some energy in these processes, typically something like 20–50 eV per event, but because it has plenty of energy (say, originally, 1 MeV), it will destructively interact with another molecule, and so on, until they lose most of their original energy. An ejected electron then acts as another β-particle. A track of photons (γ-ray) was shown in Fig. 9.2 for a lattice of carbon atoms; they are scattered randomly as they ionize, though relatively sparsely. Tracks of photons would not be exactly like this in cells and tissues, as the medium is not homogeneous. Electrons would also be scattered, but more frequently than photons, ejecting electrons as they go, and they are not as far-reaching compared to photons (Sect. 9.2.2; see also Fig. 9.3). The heavier α-particle tends to scatter electrons more frequently, as seen in Fig. 9.3, but would not travel very far.

Samples of tissues/organs from the dead bodies of Hiroshima and Nagasaki victims have been preserved (first in the United States and later returned to Japan). Recently, tracks of α-particles (likely of Pu) were observed in such a preserved sample of dead tissue of the kidney (more than half a century old), as shown in Fig. 11.5. The α-source is, indeed, still emitting radiation particles after more than half a century, as expected. This picture demonstrates that internal exposure is, indeed, a reality.

The mechanism of internal effects has not yet been delineated enough. First of all, how and where radioactive material is settled cannot be specified for each incidence, except for iodine. The material could be "dispersed" in the aqueous medium in or out of a cell, be relatively concentrated in a certain region of a cell or among cells, or exist in a very minute particle of, e.g., 1 μm in diameter. If this last scenario is the case and the minute particle is made of CsOH, then it would contain

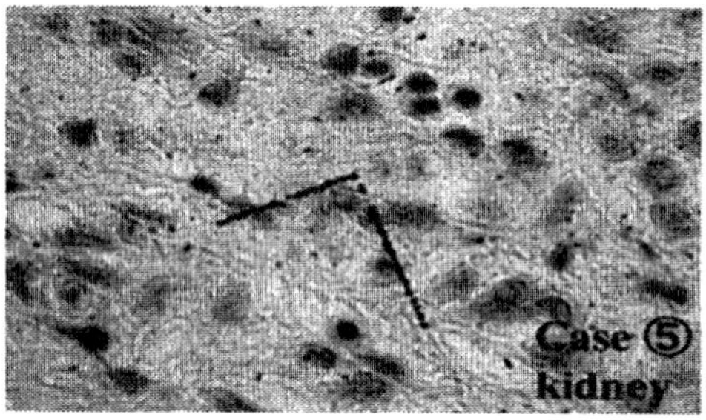

Fig. 11.5 Tracks of α-particles in the dead tissue of an atomic bomb victim (Shichijo: http://ihope.jp/2009/07/03122206.html)

ca. 1.5×10^{10} atoms of Cs-137, which would give rise to about 11 Bq. It then gives off 11 β- and γ-particles per second, and 4×10^4 of each of them over the course of an hour. If it remained there for 1 year, it would emit 3.5×10^8 particles of both β and γ. Each of these radiation particles would collide and ionize many molecules in cell after cell, until it has lost enough energy. This is schematically depicted in Fig. 11.6. In this figure, the radiation is designated α, β, and γ, but not all nuclides emit all of these particles, as shown in Table 2.1. What molecules and how many of them and how many cells are affected by a single radioactive source cannot be specified. γ-Particles may traverse more than several tens of thousands of cells, β-particles perhaps a few hundred, and α-particles a much smaller amount. But the strength of impact (damage) is in the reverse order as follows: $\alpha > \beta > \gamma$. This is expressed in terms of linear energy transfer (LET), a measure of the rate at which energy is deposited as a radioactive particle travels, with a unit of KeV/μm. LET values are given as 100 for 5 MeV α-particles, 20 for 2.5 MeV neutrons, 3.0 for diagnostic X-rays, 0.3 for 1 MeV β-particles, and 0.25 for γ-particles (Forshier 2009).

An effect called the "bystander effect" has been proposed alongside the multicellular effects of a single radiation particle, as shown in Fig. 11.6. A cell that has been hit by a radiation particle would give some damaging effects on the adjacent nonradiated cells (see Brenner 2007). For example, when all cells in a sample were hit by a single α-particle, it induced mutations on a gene CD59, and the number of such mutations was found to be something like 90 per 10^5 surviving cells. The number of mutations per 10^5 cells was about 80 when only 20 % of the cells in the sample were hit by a single α-particle. This is much higher than that expected for a situation where no hit means no mutation (cited in Brenner 2007). If so, at 20 %, the mutation rate should have been about 18 per 10^5. The results, along with similar observations, are interpreted as a "bystander effect" that an irradiated cell communicates with surrounding cells and induces some mutations in those unaffected cells. The range of the bystander effect has been estimated to be 50–100 cells around a single cell. If this effect is also taken account of, a single radioactive particle would impact a large number of molecules in a large number of cells.

What components of a cell would be affected? If it is assumed that the chance with which a radiation particle hits a specific kind of molecular entity is proportional to its mass in a cell (see Sect. 3.3; the volume instead of mass should be more

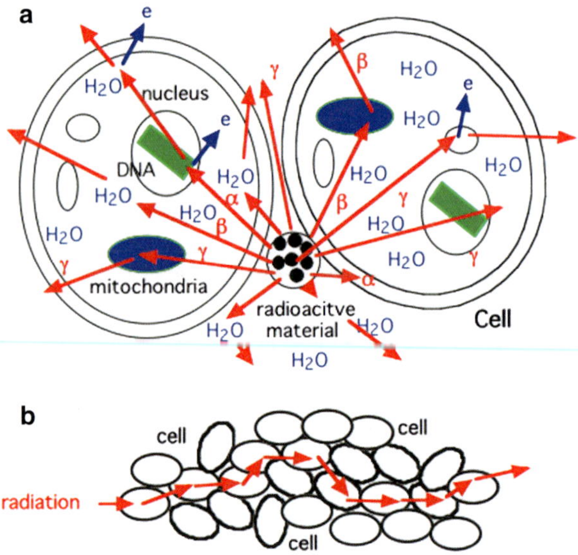

Fig. 11.6 Mechanism of internal exposure

appropriate), then the highest possibility is with water at about 0.7 (70 %, or 70 out of 100 radiation particles); the next is protein molecules at 0.18, lipids at 0.05, RNA at 0.01, and DNA at 0.003. The radiation effects on water are to produce hydroxyl free radicals and other reactive species, as discussed in Sect. 12.1. These free radicals then cause damage to other chemical entities, including DNA. The lesions on protein molecules may obstruct the proper functioning of the cell or organelle. The impact of radiation particles on membrane lipids may be more often than the mass indicates, because of the nature of its geometry (surrounding a cell), and may disrupt the integrity of the membrane, causing the leakage of cytoplasm. As will be seen in Sect. 12.1, there is no adequate defense mechanism for the damage on these molecules other than DNA. The result would be the disruption of the function of a particular organ(s), causing various diseases, and, if the disruption is great enough, may lead to death due to organ(s) failure. Examples of these effects on various organs, tissues, and cells will be discussed in Part V, in particular, Sect. 14.3.2.

The thyroid gland specifically requires iodine, as it produces thyroxine, a hormone containing iodine. Therefore, iodine (I-131 or any other iodine isotopes) tends to accumulate in the thyroid, though it would also be distributed in other parts of the body. The radiation from I-131, therefore, tends to cause abnormalities in the thyroid, which could become cancerous.

Another common radioactive material from the nuclear fission reaction is Cs-137 (134, 135, or any other Cs isotopes). It behaves like K, as they are similar in terms of chemical behavior, but it tends to stagnate in some organs and tissues, such as skeletal muscle, kidney, heart, liver, and the brain, as seen in Fig. 10.2. However, this does not necessarily mean that Cs-137 would not affect other organs and tissues.

The effects on DNA are complicated, as there are a large number of repair mechanisms for DNA involving a huge number of factors: proteins and RNAs, not necessarily for the purpose of defending against radiation in general. This issue will be detailed in the next chapter. The presence of defense mechanisms of DNA damage implies that cancers are relatively rare events and take time to appear, compared to other disease states that may be caused by radiation and would appear sooner.

11.5.5 Stochastic Nature of Low-Dose Radiation Effects

The health effects of low-level internal radiation exposure is stochastic at every stage of the process. First, whether or not a person ingests a tiny quantity of radioactive material cannot be predicted, even though they may live in a highly contaminated area. One person may receive internal radiation exposure, but their next door neighbor may not. This is mainly due to the fact that the radioactive materials distribute unevenly. As is often the case, a place a few meters away from a so-called "hotspot" may not be contaminated.

It must be noted, though, that the result of exposure can somewhat be determined individually by the so-called "whole body counter", which can determine only γ-radiation partially. Even if a small amount gets into the body, how it will distribute within the body is highly stochastic, though there is some tendency for a certain chemical entity (whether radioactive or not) to concentrate in certain tissues or organs. Once it settles in a certain tissue and radiates particles (α, β, and/or γ), the molecular components that these particles would hit are only stochastically determined. It may hit water, forming, for example, OH free radicals. It may hit certain protein molecules or a certain portion of DNA molecules or RNA, or other molecules. All these occurrences are stochastic, and the health effects caused by these radiation hits may or may not manifest. That is, the damage may be repaired before the effects manifest, or may be made null through apoptosis. But it must be pointed out that the radiation effects would persist as long as the radiation source remains in the body, even if the affected cells may be lost by apoptosis. And, hence, the same radiation source may affect other cells and molecules nearby.

Overall, it is highly unlikely to be able to determine the direct cause–effect relationship between the low-level radiation and the adverse health effect manifestation for an individual case. We will discuss some recent developments regarding this point shortly. It can be determined, though, if a systematic comparison is made between the cause of death and the radiation measurement of parts of the dead body, as will be discussed in Chap. 14. This is a verification of direct causality, though it is possible only after death.

The cause–effect relationship in the case of stochastic processes can only be surmised by epidemiological, i.e., statistical, studies. Here is the fundamental difficulty, particularly with regard to the sample size (number of persons to be studied) and a proper control group. This difficulty has been a stumbling block in establishing scientifically the direct causality in the case of low-level radiation exposure, both in the aftermath of the atomic bomb explosions in Hiroshima and Nagasaki, and the accident at Chernobyl, and, likely, the same may be said with regard to the Fukushima incident. The authorities on the side of the nuclear industry use this trick, arguing that there is no scientifically proven causality between the adverse health effects and the low-dose radiation, and that, therefore, no worrying is necessary about the low-dose exposure, implying that low-level radiation exposure has no serious health effects. However, the data accumulated so far strongly suggest a correlation between radiation and adverse health effects, as will be seen in Part V.

11.5.6 Issue of LNT and Acute Versus Protracted Exposure

There are a large number of adverse health effects possible other than cancers, as seen in this treatment. Researchers and others concerned with radiation effects have focused only on cancers, unfortunately. Debates have been ongoing regarding the effect of low-level radiation, particularly below 100 mSv (0.1 Gy). The main questions have been: Would a low-level radiation have a significant effect in causing cancer? Would a low dose give different results, if given in a single shot or protractedly? Is there any threshold dose below which no adverse effect appears? Is the relationship between the adverse health effects and the exposure dose linear or otherwise?

The debates about these issues are not very fruitful, in view of the basic mechanism of radiation effects, particularly in the case of internal radiation, the multiple factors involved, and the stochastic nature of the effects. The exposure dose used in arguing these issues is not accurately determinable, and, besides, the major trend in dealing with the issue is to ignore the internal radiation, or it does not distinguish the internal from the external radiation. If it can be assumed that the dose values used are consistently evaluated, the relative Gy or Sv values may be meaningfully compared among them. It may be worthwhile to argue whether cancer risks are correlated with the exposure dose, if the cancers caused by radiation are statistically consistent among all the individuals. Under the assumption that this is the case, we may argue whether the relationship between the cancer extra relative risk (ERR, see Chap. 13) and the dose is linear without a threshold or nonlinear with a threshold.

These issues were reviewed and discussed fairly well by J. Beyea (2012). First, the recent study on atomic survivors, i.e., LSS-14 (Ozasa et al. 2012) clearly showed that a linear no-threshold (LNT) model holds for a wide range of survivor doses, even below 100 mGy (Sect. 13.2.3, Fig. 13.5). Many other studies have also indicated the LNT model. Some authors also point out that delayed cancers are the usual focus of the low-level radiation debate, but delayed stroke and heart disease account for "about one third as many radiation-associated excess deaths as do cancers among atomic bomb survivors" (Shimizu et al. 2010, cited in Beyea 2012).

Elaine Cardis and her colleagues (2005) surveyed 400,000 nuclear industry workers in 15 countries who had been chronically exposed and whose doses had been recorded. The study revealed a higher incidence for protracted exposure than that found in the atomic bomb data (one shot of high exposure). This implies that protracted exposure shows a higher cancer risk than the one-time exposure of the same dose, and that the radiation effect is cumulative, implying that the remedial mechanisms are not sufficient to nullify the slow and protracted exposures, even at low levels.

References

Beyea J (2012) The scientific jigsaw puzzle: fitting the pieces of the low-level radiation debate. Bull Atom Sci 68:13–28

Brenner DJ (2007) Radiobiology for radiation epidemiologists. NCI REB Radiation Epidemiology Course, May 14–17

Caldicott H (2011) Nuclear power is not the answer. The New Press, New York

Cardis E, Vrijheid M, Blettner M et al (2005) Risk of cancer after low doses of ionising radiation: retrospective cohort study in 15 countries. Br Med J 331:77

DS02 (2009): see Young et al. (2003)

ECCR (2010) Recommendations of the European Committee on radiation risk –the health effects of exposure to low doses of ionizing radiation

Forshier S (2009) Essentials of radiation biology and protection, 2nd edn. Delmar, Clifton Park

Garger EK, Kashpur V, Paretzke HG, Tschiersch J (1998) Measurement of resuspended aerosol in the Chernobyl area. Part II: size distribution of radioactive particles. Radiat Environ Biophys 36:275–283

Kaneyasu N, Ohashi H, Suzuki F, Okuda T, Ikemori F (2012) Sulfate aerosol as a potential transport medium of radiocesium from the Fukushima nuclear accident. Environ Sci Technol 46:5720–5726

Kendal G, Little MP, Wakeford R, Bunch KJ, Miles JCH, Vincent TJ, Meara JR, Murphy MFG (2012) A record-based case-control study of natural background radiation and the incidence of childhood leukaemia and other cancers in Great Britain during 1980–2006. Leukemia 27 (2013):3–9 (Published online 6 June 2012)

LSS-14: see Ozasa et al. (2012)

Lupandin VM (1998) http://www.rri.kyoto-u.ac.jp/NSRG/reports/1998/kr-21/Lupandin.html

Ozasa K, Shimizu Y, Suyama A, Kasagi F, Soda M, Grant EJ, Sakata R, Sugiyama H, Kodama K (2012) Studies of the mortality of atomic bomb survivors, Report 14, 1950–2003: An overview of cancer and noncancer diseases (LSS-14). Rad Res 177:229–243

Stewart A, Webb J, Hewitt D (1958) A survey of childhood malignancies. Br Med J 1:1495–1508

Young R, Egbert RW, Cullings SD, Kerr HM, Cullings GD, Imanaka T (2003) Survivor Dosimetry, Chapter 12. In: A new dosimetry for the atomic-bomb survivors, DS02. RERF

Chapter 12
Control of Cell Processes and Defenses Against Radiation

12.1 Mechanisms to Defend Against Damaging Effects or to Repair Lesions on Non-DNA Molecules

Living organisms are always subject to disturbing/damaging effects, both endogenous and exogenous. They have had to develop ways to defend against such effects; otherwise, they would not have survived. No comprehensive treatment of this issue is intended here, and an emphasis is placed on radiation.

12.1.1 Effects of Nonradioactive Entities

12.1.1.1 Effects of Exogenous Factors

First, there is usually a redundancy for biocompounds, except, perhaps, for DNA in a cell, as estimated in Sect. 3.3. Even DNA is double-stranded, and is, of course, a kind of redundancy. Hence, damage to a single molecule would not matter very much (except for DNA), and the cell may continue to function undisturbed, except for such molecules whose redundancies are low and very critical for the function of the cell. Besides, it is known that proteins incorrectly formed in cells are swiftly degraded by some enzymes, because the accumulation of malformed proteins tends to aggregate and disrupt the cellular functions. It does not seem, though, that this function works for the kind of lesions caused by radiation.

Toxic substances have various characteristics and cannot be generalized. Just a few examples will be mentioned. Antidotes and defense mechanisms usually exist for most chemical poisons. Carbon monoxide binds to the iron of hemoglobin more strongly than dioxygen, thus, resulting in asphyxiation. No adequate antidote for the effects of CO exists. Cyanide is another common poison, which blocks, among others, the last step of the respiratory process. Many heavy metals such as mercury (Hg), lead (Pb), and cadmium (Cd) as well as arsenic (As) are common poisons.

E. Ochiai, *Hiroshima to Fukushima*, Science Policy Reports,
DOI 10.1007/978-3-642-38727-2_12,

Defense mechanisms exist in some organisms for some of these poisons. For example, defense mechanisms against Hg include ways to bind it strongly (metallothionein or binding with Se) and to convert mercury compounds to volatile forms (reduce it to mercury metal or convert it to dimethylmercury) (Chap. 11, Ochiai 2008).

Biological toxins such as frog toxins and snake venoms attack specific tissues/organs. As long as the symptoms are recognized early enough, effective antidotes can usually be applied to nullify the effects. There are many other artificially produced chemicals which have entered into the environment and can then get into living organisms.

Naturally occurring chemical entities (including those which are biologically derived) that cause negative health effects as exemplified above are usually avoidable if one is careful, and counter-measures by chemistry to reduce the effects are usually available. However, the human species has invented and produced in recent centuries an enormous number of chemicals which are artificial and new to living organisms. The organisms may tolerate some of them or may not be able to tolerate others. Some organisms may develop mechanisms to deal with new chemicals in several later generations. Some chemicals may be such that defending mechanisms are difficult to develop. These are the chemicals that are causing environmental problems.

12.1.1.2 Control of the Quality of Proteins

The proteins produced by the DNA-RNA ribosome system may or may not be proper in some respects. Abnormal proteins would disrupt the cell functions, and, hence, there are a few remedies to control the quality of proteins. There are four levels in the protein structures: the primary structure is the amino acid sequence (of a polypeptide), the secondary structure is the α-helix, β-strand, or a random coil of polypeptide, the tertiary structure is the folded three-dimensional structure, and the assemblage of subunits is the quaternary structure. The information contained in DNA determines only the primary structure. In many cases, the primary structure can dictate the secondary and tertiary structures correctly, and, hence, no help or control is necessary for those proteins. Some proteins may need some help to fold themselves in proper ways.

Most of the misfolded proteins have more hydrophobic residues on the surface. They would coagulate, forming precipitates in the cytoplasm, thus, disrupting the cell functions and could cause diseases. Specific proteins called "chaperones" help proteins fold properly. Two such proteins are heat shock proteins; one of them acts while the protein is still in the ribosome, and the other acts after the protein comes out in the cytoplasm. A protein that has failed to be corrected in the tertiary structure via chaperones would then face decomposition by the proteasome. A misfolded protein is first recognized and marked by ubiquitin. The tagged protein is then subject to the first portion of the proteasome, which unfolds the protein, and is then translocated to the second portion. It is essentially proteases, protein-hydrolyzing enzymes, and decomposes the polynucleotide by hydrolysis (see, for example, p. 355–363, Alberts et al. 2002).

Protein decompositions take place under normal conditions as well, as some proteins need to be eliminated quickly after they have performed their functions. The proteasome is also used for this purpose. Therefore, proteases are abound in a normal cell, amounting to as much 1 % of all proteins in a cell.

12.1.2 Remedies for Radiation Effects

Radiation is quite different from the effects by chemicals as mentioned above. First, radiation is unavoidable; the only way to avoid it is to live in a completely radiation-free environment, where everything including air, food, and accessible water are uncontaminated. The problem is that this is not easily achievable because radiation is invisible in many senses. Secondly, no counter-measure is available to negate the radiation effects. However, there are some remedial measures present in living systems that repair the damage caused by radiation. This issue is discussed below.

It must be pointed out at the outset that remedial measures are limited particularly to non-DNA biochemicals, and defense mechanisms for DNA, though elaborate, would not be able to completely negate the radiation effects. This situation will be reflected in the health effects of radiation. In other words, noncancerous diseases will manifest immediately in the people who are exposed, but there is some delay in the emergence of cancers in general, though some cancers such as thyroid gland cancer may appear fairly soon after exposure. Noncancerous diseases are many and varied, but they have not been recognized as such, i.e., radiation-caused, partly because numerous causes for such diseases are possible.

12.1.2.1 Indirect Effects

The most commonly acknowledged function of ionizing radiation is the formation of free radicals, particularly due to the decomposition of water molecules or the addition of electrons to dissolved O_2 molecules, as follows (Ochiai 1994):

$H_2O + h\nu$ (represents both photons and particles) $\rightarrow H_2O\bullet^+ + e$ (electron)
$H_2O\bullet^+ + H_2O \rightarrow \bullet OH + H_3O^+$
$H_2O + h\nu \rightarrow H_2O^\#$ (excited state) $\rightarrow H\bullet + \bullet OH$ (hydroxyl radical) (bond breakage)
$O_2 + e$ (β or other) $\rightarrow \bullet O_2^-$ (superoxide radical)

The "•" beside an entity represents an odd (unpaired) electron, and an entity with a "•" beside it is a free radical. As about 70–80 % of the cell content is water, the formation of free radicals from it and O_2 is the most likely event caused by radiation. A free radical is very reactive, and will react with a first encounter, by removing, for example, a hydrogen atom from it, making it another free radical, which reacts further. For example:

$R–H + \bullet OH \rightarrow R\bullet + H_2O$
$R\bullet + O_2 \rightarrow ROO\bullet$
$ROO\bullet + H\text{-}R' \rightarrow ROOH + \bullet R'$
$\bullet O_2^- + H_2O \rightarrow HOO\bullet + {}^-OH$
$R–H + \bullet OOH \rightarrow R\bullet + HOOH\ (H_2O_2)$
$HO\bullet + \bullet OH \rightarrow HOOH$ etc.

The chemical entity RH in the equations above can be anything, including DNA, RNA, lipids, steroids, carbohydrates, and enzyme proteins. If it is a membrane lipid, the integrity of the membrane may be disrupted. An enzyme may become dysfunctional as a result of these reactions. The effects on DNA and RNA will be discussed later.

These entities containing O atoms, i.e., HOOH (hydrogen peroxide), ROOH (hydroperoxide), O_2^-, and OH, are reactive, more so than the ordinary oxygen O_2, and, hence, are collectively called "reactive oxygen species" (ROS). Most of these toxic entities except for the hydroxyl radical (•OH) would form indigenously under normal physiological conditions, and, hence, many biological systems have developed some mechanisms (enzymes) to detoxify them. Examples include (Chaps. 6, 7, and 9, Ochiai 2008):

Catalase: $2HOOH \rightarrow 2H_2O + O_2$
Peroxidases: $ROOH + AH_2 \rightarrow ROH + H_2O + A$
Glutathione peroxidase: 2GS-H (glutathione) + ROOH $\rightarrow$ GS-SG + ROH + H_2O
Superoxide dismutase: $2\bullet O_2^- + 2H^+ \rightarrow O_2 + HOOH$

In these equations, AH_2 represents a reducing agent that can provide two Hs (or two electrons); examples include pyrogallol, ascorbic acid (vitamin C), and NAH.

However, it must be pointed out that there are other reactions that exacerbate the free radical effects. For example, any Fe or Cu ions present in the cell can form free radicals from hydroperoxides (ROOH); thus:

$ROOH + Fe(II) \rightarrow RO\bullet + {}^-OH + Fe(III)$
$(RO\bullet + Fe(II) + H^+ \rightarrow ROH + Fe(III))$
$ROOH + Fe(III) \rightarrow ROO\bullet + H^+ + Fe(II)$

It is known that the presence of an excess of Fe can exacerbates "oxidative stress". It can be inferred from these various reactions that no effective enzyme has been developed for detoxifying the free radicals, except for superoxide radicals, perhaps because these radicals are too reactive to adequately deal with, but usually of a transient nature.

There are, however, a number of biocompounds that deactivate free radicals, including •OH radicals. Those compounds are good reducing agents, and would become stable as a result of donating an electron or two. Most of them are known as antioxidants. Examples include ascorbic acid (vitamin C), polyphenols, thiols such as glutathione, and compounds of the flavin type. Each of these compounds reacts with a free radical, typically giving an electron (and H^+) to it and nullifying its free radical character. As a result, it, by itself, would become a free radical, but this free radical is relatively stable, and not very harmful. Yet, it can further react with another free radical to nullify the free radical character.

Just a few examples are shown below. X with a dot represents any free radical, and it abstracts a hydrogen atom from these chemicals (dihydrogen flavin FH_2, ascorbic acid, polyphenol) to become non-free radical XH. The intermediate entity in these reactions is relatively stable, and then allows another free radical to abstract a hydrogen atom, resulting in a stable compound as shown.

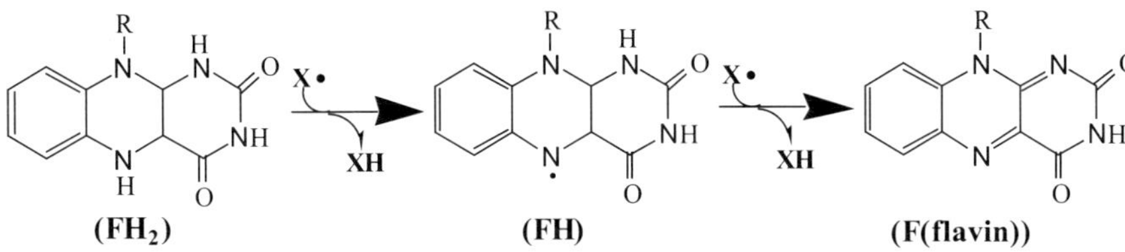

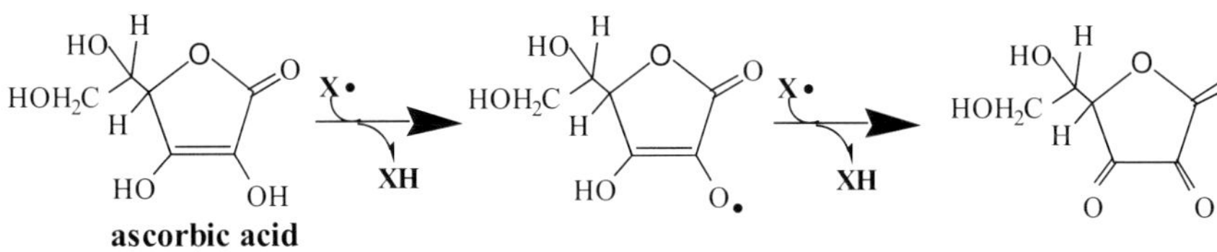

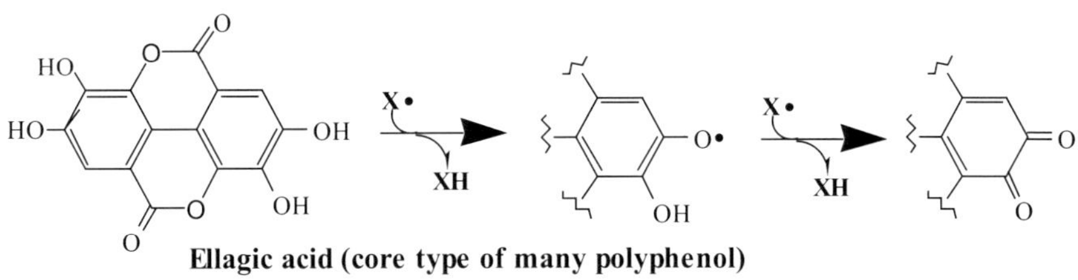

[**Note**: O_2 itself is a free radical, but it is a special one (biradical, triplet) and not as reactive as regular (doublet) free radicals].

12.1.2.2 Direct Effects: No Defense

Radioactive particles can and will attack molecules other than water or DNA, and ionize them or break bonds within them. How the lesions on the biomolecules caused by these attacks would affect the health of an entire body has not been intensively studied. Most molecules are present redundantly in a cell. As many as thousands to tens of thousands of copies of a single enzyme protein are present in a cell, for example. Therefore, no significant effect may manifest unless a substantial number of copies of it were destroyed. However, there are biomolecules for which not many copies are present to begin with, and damage to such molecules can disrupt the function of a cell. This is true with most of the cell signaling proteins. They exist normally in a few copies, and upon the arrival of a signal, they will be multiplied through a cascade process to exert their effects. If such a signal protein is damaged, no alternative is available and a serious disruption may result. The mechanisms to control the quality of proteins as mentioned in Sect. 12.1.1.2 are unlikely to help restore the damage on proteins caused by radiation or decompose such a protein.

The cell membrane would inevitably be subject to radiation effects, due to its geometrical nature. Zapping through it by ionizing radiation may disrupt the integrity of the membrane and break it, resulting in the release of the cell content. Most organelles (including mitochondrion), with lipid membranes, may also be subject to damage in their membranes, which may result in the breakage of the organelles.

Mitochondrion contains an assembly of enzymes and proteins involved in the TCA cycle and the electron transport system, as well as DNA (separate from the nuclear DNA). Ionizing radiation on mitochondrion can disrupt the production of ATP, the energy carrier for life, and could lead to cell death. It seems that the electron transport system is particularly vulnerable to ionizing radiation, because it can eject the electrons moving among proteins. The ribosome is where proteins are synthesized, and its destruction diminishes proper protein syntheses.

Which tissues or organs would be subject to radiation effects depends on what chemical species are responsible for the radiation, because the behaviors in biological systems of the radioactive material are governed by their chemical properties, not whether they are radioactive or not. The cases of Cs and I are the most intensively studied, and were dealt with in Sect. 10.2, and will be discussed in Part V.

12.2 Control of Cellular Processes

12.2.1 Organism/Organs/Tissues/Cells

A multicellular organism consists of organs/tissues/cells/organelles/molecules, all of which are intimately related to each other in order to function properly (healthily). The physiological manifestations of radiation effects are related to all of the body's components. Two crucial molecular entities of a cell are "proteins" and "RNA/DNA". Some of the essential cell components (though there is no nonessential component in a cell) are the nucleus, mitochondria, ribosomes, and cell membranes. The radiation effects on proteins, mitochondria, ribosomes, and cell membranes were briefly outlined in the previous section. Not much detail has yet been unraveled with respect to radiation effects on these entities, though some observations have been reported, as will be outlined in Sect. 14.3.2 (Bandazhevsky 2000). However, studies on DNA are making rapid progress, and producing an enormous amount of literature. It is not our intention nor is it possible to provide the complete details of the mechanisms of DNA repair and cancer formation. Only a brief outline will be given, in order to see how radiation might be involved in tumor formation.

Here is an outline of the entire system at the level of a cell and its adjacent cells.

The cell in question is at a certain stage of the cell cycle (see the next section). It is in contact with several cells, which might give certain signals to the cell in question. It also receives a number of chemicals, including hormones, drug chemicals, and others, including radioactive chemicals. DNA might be in the process of being transcribed into an mRNA, involving enzymes including RNA polymerase. An mRNA would then be translated into a protein in the ribosome, where a number of small RNA (t-RNAs) and large catalytic RNA are working. Transcription and translation are all controlled by a large number of proteins, which are the products of other areas of the DNA. DNA might be in the process of replication, which also involves a large number of enzymes/proteins, which may obstruct or enhance the replication, or mend the wrongly attached nucleotides, etc. At this cellular level, the living process involves a large number of entities (proteins and nucleic acids), and their interactions are complicated.

It seems easy to disrupt such a complex system, but, remarkably, it maintains normality fairly well. How it is done is briefly discussed in the next few sections. The question is whether the system is sturdy enough to prevent the adverse effects of radiation. The effects of radiation will be exerted directly or indirectly on any of these molecular entities, proteins, DNA, and RNA. The researches on cancer-causing radiation effects tend to be focused on damage to DNA. It may or may not be a major factor, and there are many other avenues that need to be explored. We will focus on DNA not only because it is the most important entity, but also because no other component such as proteins has been studied adequately.

12.2.2 Cell Cycle

There are two different cell types in living organisms: one is "somatic" and the other is "germ". The somatic cells contain two copies of all double-helix DNA in 22 pairs and X and Y chromosomes in humans, while a germ cell contains a single copy each of all double-helix DNA. The somatic cells proliferate during the growing period (embryonic, fetus, and childhood periods), and those in certain tissues keep proliferating throughout a person's life, as in skin, the epithelial cells of the gastrointestinal tract, and blood. Others stop multiplying, like those in the central nervous system and skeletal muscle.

The cells that divide and proliferate go through a cyclic process termed the "cell cycle". The main events are the replication of DNA (S-phase) and the division of replicated chromosomes into two halves (mitosis; M-phase). After M-phase, the cell divides into two daughter cells (cytokinesis). In between the M- and S-phases are two intervening phases, termed as G_1 and G_2, as shown in Fig. 12.1. The

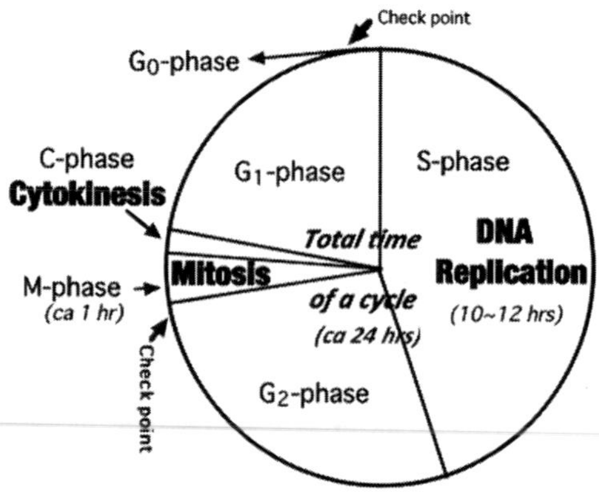

Fig. 12.1 Cell cycle

manufacturing of other cellular components: proteins, organelles such as mitochondria, etc. are carried out during the $G_1/S/G_2$ phase. In a typical proliferating cell in the human body, a single cycle takes about 24 h, though the cell cycle time varies from one tissue to another, ranging from hours to days. A static cell remains in the G_0-phase. The operation of a cell cycle is usually subject to many disturbing factors; exogenous (including signals from adjacent cells) and endogenous. Therefore, there are check and control mechanisms in order to ensure proper cell cycle processes. Two checkpoints have been recognized; one at the G_1–S transition point and the other at the G_2–M transition point (as indicated in Fig. 12.1).

The cycle is controlled mainly by a set consisting or proteins called cyclins and associated proteins. A cyclin binds with and activates a cyclin-dependent kinase (CDK). CDK thus activated phosphorylates proteins/enzymes at serine and/or threonine residues, initiating a cascade of processes to start the synthesis of DNA at the junction of the G_1–S phase, for example. This is carried out by the cyclin 2E-CDK2 combination. Cyclin A-CDK2 works for the S-phase and cyclin B-CDK1 (CDC2) works for the G_2–M checkpoint. The full activation of CDKs requires a further action of another enzyme, CDK-activating kinase (CAK). Cyclins are synthesized and decomposed as the cell progresses in its cycle. Many somatic cells would remain in the G_0-phase, where the cell cycle activity is absent, until such a time that the activation of the cells becomes required. Brain and skeletal muscle cells, once formed, would remain in the G_0-state.

How, then, would DNA replication be initiated? It begins at the "origin of replication"; such sites are scattered at various locations on chromosomes. It is recognized and bound by a large protein, origin recognition complex (ORC). Another protein, cell division cycle 6 (CDC6), binds to ORC; to this site binds minichromosome maintenance (MCM) protein, surrounding the double-helix DNA. Cyclin 2E-CDK2 mentioned above then phosphorylates and degrades CDC6, and phosphorylates ORC, and then replication begins with DNA polymerase, helicase, and the accessory proteins and enzymes. These are factors present in a cell nucleus. The cell cycle control is summarized in Fig. 12.2, in which the arrows

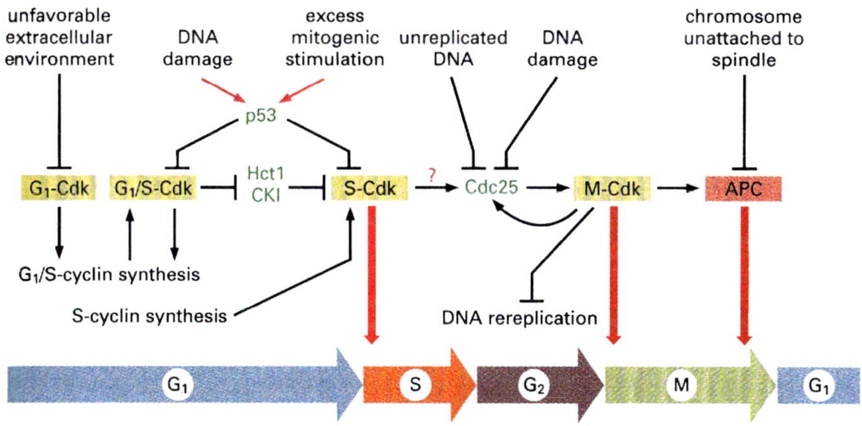

Fig. 12.2 Main events in cell cycle control (Fig. 17.34 in Alberts et al. 2002)

indicate promotion, including both black and red lines, and lines with a bar at the end indicate inhibition. So, for example, unreplicated DNA inhibits Cdc25, hence, Cdc cannot activate M-Cdk, and, as a result, no mitosis would start.

A cell receives inputs from outside on its receptors in the membrane, which regulate the cell cycle. For example, fibroblast cells have a receptor for platelet-derived growth factor (PDGF). When PDGF binds the receptor, the receptor is activated as tyrosine kinase phosphorylase, and activates the *ras* gene (the RAS protein is a GTPase), which initiates a cascade of multiplication of signal proteins, which enter the nucleus and eventually override the cellular control that otherwise inhibits cell division.

Further details need to be referred to in an enormously voluminous book, such as "Molecular Biology of the Cell" (Alberts et al. 2002, 2007). This short description is given just to indicate how complicated and how many factors (proteins) are involved in cell cycle control and proper operation of a cell. All these factors, in addition to the DNA itself (though they are present in chromosomes, double-strand helices, and other forms in cells), are potentially subject to radiation effects.

12.3 Damage to DNA and Controlling Systems

DNA molecules are constantly subject to altering/damaging effects by endogenous reactive metabolites, therapeutic drugs, environmental mutagens, and others. However, DNA needs to be kept as intact as possible, because it is the critical entity in living systems. Living systems have developed very elaborate mechanisms to maintain the structural integrity of DNA/chromosomes. In addition, a cell itself may be destroyed (apoptosis) if the repair of damaged DNA was insufficient, so that the damage effect on DNA is nullified.

12.3.1 Damage/Modifications Caused by Nonradioactive Chemicals

12.3.1.1 On DNA

DNA is a linear polymer of deoxyribonucleotides. Deoxyribonucleotides consist of three components: deoxyribose, base, and phosphate. As for base, there are two purines: adenine (A) and guanine (G), and two pyrimidines: cytosine (C) and thymine (T). They are shown schematically in Fig. 12.3 (modified from Alberts et al. 2002). This is a portion of a single strand. DNA in a cell (in both prokaryotes and eukaryotes) is double-stranded through hydrogen bonds between A and T, and G and C, normally, and the double strand coils itself in a helix. In eukaryotes, it further wraps around the protein histone, forming nucleosomes, and chromatin, the chain of nucleosomes, which is then further condensed into chromosomes.

The arrows show the sites where chemical changes can take place, and the thickness of the arrow represents the degree of vulnerability. The blue arrows show hydrolytic changes, the red arrows indicate oxidative changes, and the green arrows show where methylation takes place. Numbers (1) – (9) indicate the types of chemical reaction. These changes are brought about by many causes: chemicals, temperature fluctuations, metabolic accidents, exposure to some environmental substances, and various sorts of radiation. Radiation may cause many more different changes: bond breakage, ionization, etc., but a discussion of this issue will be deferred to the next section.

Reaction (1) removes a base and, hence, disrupts the association between two strands. It has been estimated that about 5,000 purine bases are lost from the DNA in every cell of the human body per day (p. 268 in Alberts et al. 2002). Reaction (2) is called deamination and changes, for example, cytosine to uracil, which prefers adenine instead of guanine in hydrogen bonding. About 100 cytosine-to-uracil conversions take place in every cell in a day (p. 268, Alberts et al. 2002). Reaction (3) cuts the single strand of DNA (SSB = single-strand breakage) and may be carried out by various nucleases. The reaction of this type by a restrictive enzyme endonuclease is to destroy the DNA of an invading virus. Methylation, reactions (4) and (5) removes the hydrogen bonding ability of the methylated base. Methylation at C-5 of cytosine is different in its effect, and has a lot to do with gene control. The structures resulting from reactions (6) and (7) may jeopardize the recognition of a specific sequence in the DNA helix, as the regulatory proteins search for such a site. Reactions (8) and (9) would be highly unlikely under ordinary circumstances, but would disrupt the double-helix structure if they take place. The two strands may be cleaved simultaneously at the same or different locations: "double strand breakage" (DSB). This can happen most often from radiation effects.

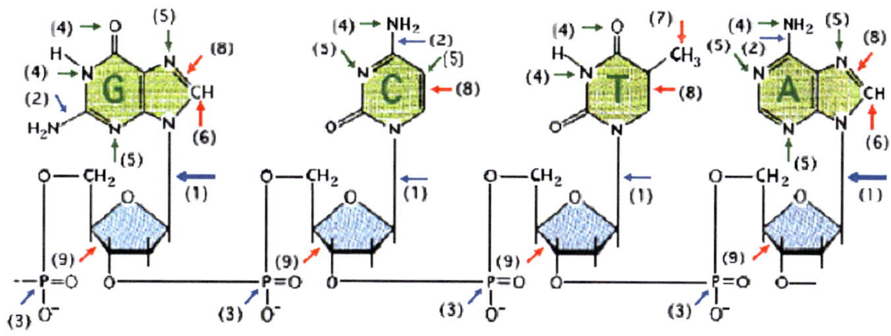

Fig. 12.3 Modifications that DNA may undergo (Modified from Fig. 5.46 on p. 268 in Alberts et al. 2002)

12.3.1.2 Modifications of Bases of DNA and Histone: Epigenetic Effects

The expression of a specific gene is controlled not only by the DNA sequence but also by many controlling factors, signal proteins, and also modifications of bases in DNA (by methylation and others), and posttranslational modification of the histone protein (by acetylation, phosphorylation, methylation, ubiquitinylation, and sumoylation). The expression of a specific gene may be altered by these factors without altering the sequence in the DNA itself. This is termed "epigenetic effects".

Histone protein consists of eight monomeric polypeptides, and the N-terminal of each polypeptide sticks out as a tail from the histone body. When some residues of the tails are modified, the conformations alter and affect the wrapping of DNA chains, thus, changing the chromatin conformation. Such modifications occur with acetylation or methylation of lysine residue and phosphorylation of serine residue; these reactions are carried out by specific enzymes. The positive charge on lysine ($-NH_3^+$) will be lost when it is acetylated or methylated. The neutral serine OH will become negatively charged when it is phosphorylated. These modifications would change the electric charge distribution and, hence, affect the electrostatic interactions between the histone and the negative charge of the phosphate bridge on DNA. The conformation changes by these modifications will result in transcriptionally active or inactive states of chromatin.

DNA is usually significantly methylated. Methylation occurs at the C-5 position of cytosine in CG sequences. Shortly after fertilization, most of the methyl groups on DNA are eliminated by demethylase. Then, in the process of development, new patterns of methylation are established. The methylation reaction is carried out by an enzyme DNA methyltransferase (DNMT). This pattern (of methylation) is to be transmitted to daughter cells upon division. This is done with the help of an enzyme, maintenance DNA methyltransferase; this enzyme methylates preferentially the CG sequences in the daughter DNA using the parental methylated DNA as the template. Hence, the methylated pattern is transmitted to the daughter DNA. It turned out that methylated segments are found in transcriptionally silent regions.

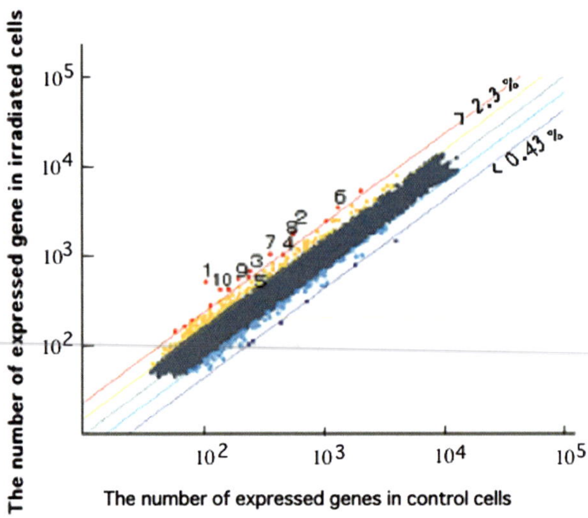

Fig. 12.4 The effects of low-dose radiation on gene expression (Fujimori et al. 2005)

That is, methylation plays a silencing role on segments of DNA, contributes to the stability of DNA, and, hence, controls development.

12.3.2 Lesions on DNA Caused by Radiation

12.3.2.1 Gene Expression Under Low-Dose Radiation

A study of the overall effect of a low dose of radiation on gene expression in human cells was carried out using human fibroblast HFLIII cells (Fujimori et al. 2005), using a comprehensive expression analysis method (based on extracted m-RNAs). Cultured cells were irradiated with 10 mGy of X-rays over 10 s, and then analyzed after 1 and 2 h. About 23,000 genes were counted in this analysis. Figure 12.4 shows the number of genes expressed in the irradiated cells versus those in the corresponding unirradiated cells 1 h after the irradiation. The majority of the genes were expressed approximately the same in both groups, i.e., along the diagonal; perhaps these genes happened to be unaffected by the radiation or were repaired enough within 1 h. Some genes were expressed up to two to three times more in the irradiated group, whereas some others were expressed less (down to less than a half). Some of the overexpressed genes are numbered in the figure, and were identified. Three of those were CXC chemokines and their related genes. The underexpressed genes were not studied. It appears that different genes would react differently to irradiation. However, the mechanism of these different reactions to radiation by the different genes have not been studied.

12.3.2.2 Gene Mutations, Chromosome Aberrations, and Effects on Epigenesis

Most of the chemical reactions mentioned in the previous section (Fig. 12.3) take place only in specific sequences, locations, and chromosomes, but radiation effects would not choose the chromosomes, the associated proteins, or the locations, etc., and the effects would be random and severe. All the reactions depicted in Fig. 12.3 could be caused by radiation as well, i.e., the bond cleavages of all types, including reactions (8) and (9), without the use of enzymes, water, or exogenous entities. Besides, the bonds that are not affected by chemical reactions (1)–(9) would also be subject to radiation effects. If such damage cannot be repaired by the existing DNA repair mechanisms, it may lead to mutations, which may eventually end up as tumors.

Ultraviolet light can excite the C = C double bond in pyrimidines (cytosine and thymine) and dimerize two pyrimidine bases on the adjacent position, as illustrated in Fig. 12.5. A similar dimerization can occur between C and C or C and T as well. This reaction is specific to ultraviolet light; radiation of higher energy would destroy a molecule, rather than electronically excite a segment in a molecule. How pyrimidine dimers in DNA cause tumors is illustrated by work on fish thyroid tumors by Hart et al. (cited in Jorgensen 2007).

Chromosomes may be cut (DSB) and recombined in the original way (homologous end joining), but also in slightly different ways from the original (nonhomologous end joining), or sometimes recombining with another chromosome (translocation), resulting in chromosome aberrations. The chromosome aberrations or abnormalities may be seen under a microscope. An example is shown in Fig. 12.6, which shows the chromosomes in a breast tumor cell in metaphase during mitosis; they are artificially stained with different colors. In this particular example, the aberrations were not necessarily caused by radiation. Here, 48 chromosomes are present instead of the normal 46, and chromosomes consisting of segments of different colors are the result of recombinations of segments of originally different chromosomes. For example, the chromosome indicated by a white arrow consists of two pieces of chromosome 8 and one piece of chromosome 17. The chromosome aberration can be caused chemically (and biologically), as well as by radiation.

Another anomaly is micronuclei. They are small extranuclear bodies, which form during mitosis as a result of parts of chromosomes or chromatids being broken off. This can be caused by radiation or chemicals.

12.3.2.3 Damage to DNA: a Summary

The phenomena observed in the sections above suggest that significant damage on DNA molecules may be caused by radiation. The possible damages are: strand beaks (single-strand break (SSB), double-strand break (DSB)), base damages (excision, oxidative damage, alkylation, bulk adducts), and crosslinks (interstrand, intrastrand (e.g., Fig. 12.5), and DNA protein). Most of these damages are common

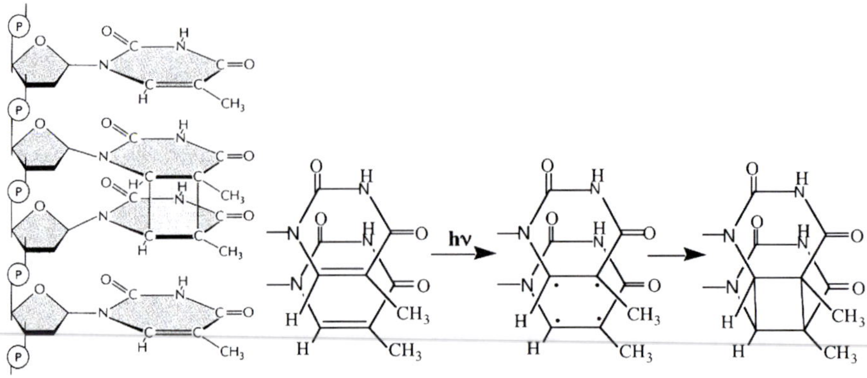

Fig. 12.5 Dimerization of thymine by ultraviolet radiation

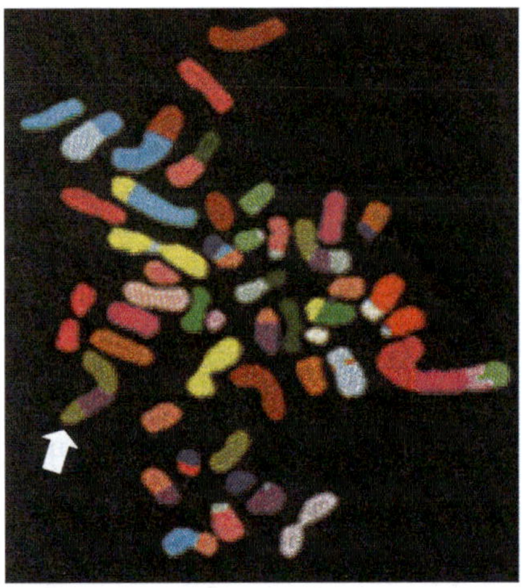

Fig. 12.6 Chromosome aberrations (Fig. 23.12, p. 1321 in Alberts et al. 2002)

to the damage done by chemicals and biological agents with a few exceptions, as described above.

12.3.3 Cell Cycle Control upon DNA Damage

12.3.3.1 Three Options When DNA Is Damaged

When DNA is damaged, there are three options for the cell to take. One is to repair the damaged DNA. The second is to arrest the cell cycle at the G_1-phase long enough for the repair mechanism to complete its role. The third option is to destroy itself (apoptosis) when the first two options have failed. This scheme is depicted in

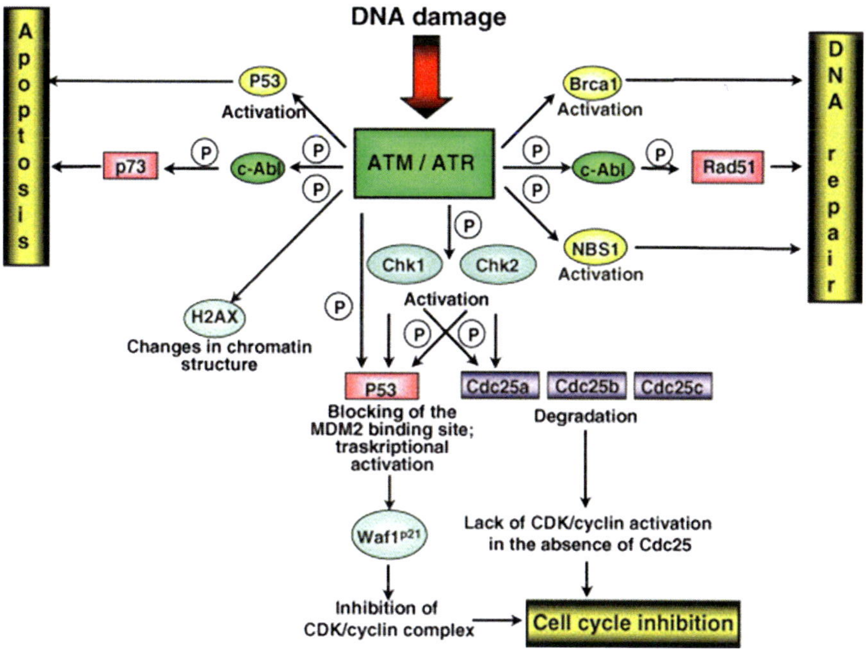

Fig. 12.7 Three options for combating damage on DNA (Fig. 6 in Christmann et al. 2003)

Fig. 12.7 (Christmann et al. 2003). The circled P implies phosphorylation (addition of a phosphate group) by kinases. If all three options fail, the damaged DNA will be replicated as such, and will be propagated with cell division. This results in mutations of certain genes.

12.3.3.2 Arrest of Cell Cycle

The damage on DNA must be detected to begin with. And that information on the damage needs to be relayed to the mechanisms of these three operations. The recognition of DNA breaks is accomplished by a group of phosphatidylinositol-3-kinases (PI3K): ATR (ataxia telangiectasia related), ATM (ataxia telangiectasia mutated), and the catalytic subunit of DNA-PK (protein kinase). ATM or ATR binds to broken DNA ends and possibly some DNA adducts. Its intrinsic activity is then activated and phosphorylates Chk1 (checkpoint kinase: in the case of SSB) in response to ATR kinase or Chk2 (in the case of DSB) in response to ATM kinase. The phosphorylated Chk1 or Chk2 then phosphorylates Cdc 25 (cell division cycle) phosphatase. This results in the degradation of Cdc25. Phosphatase is an enzyme to remove phosphate group instead of adding phosphate group (by kinase), to activate CDK/cyclin, but this activation cannot take place because of the loss of Cdc25 and, hence, the cell cycle will be inhibited. This is one way in which cell cycle arrest is accomplished when DNA is damaged. Another way to arrest the cell cycle is through the p53 protein, as shown in Fig. 12.7. Murine double minute 2 (MDM2) is ubiquitin ligase and recognizes and inhibits p53 activation, and hence, it acts as an oncogene. p21/Waf1 is a cyclin-dependent kinase inhibitor which inhibits CDK/cyclin, and, hence, arrests the cell cycle at the G_1–S checkpoint.

12.3.3.3 Apoptosis

Apoptosis is a programmed cell death (suicide), and a normal process in the development of animals. The relationship between DNA damage and apoptosis was reviewed by Norbury and Zhivotovsky (2004). ATM/ATR bound with damaged DNA directly activates p53 protein or through Chk2 as shown above, both by phosphorylation. The activated p53 then facilitates the transcription of genes encoding proteins/enzymes involved in caspases (cysteine aspartate proteases), which decompose many proteins in the cell, leading to its death. Alternatively, p53 acts on the control factors of mitochondrial membrane permeability, and increases its permeability, releasing cytochrome c, and, hence, destroys the energy-producing mitochondria.

There are routes independent of p53. ATM phosphorylates c-Abl tyrosine kinase, which then phosphorylates p73 protein, which then leads to activation of the apoptosis mechanism on mitochondria permeability. Chk2 activated as mentioned above leads to the activation of a transcription factor E2F1, which then activates genes to produce caspases. It must be pointed out, though, that not all the details of these processes have yet been unequivocally delineated.

12.3.4 Repair of Damaged DNA

A number of DNA-repairing mechanisms are operating within cells. They include nonhomologous end joining (NHEJ), base excision repair (BER), nucleotide excision repair (NER), homologous recombination repair (HRR), illegitimate recombination repair (ILRR), mismatch repair (MMR), and double-strand break repair (DSBR). These are reviewed extensively by Christmann et al. (2003) and more recently by Thompson (2012). The review by Christmann et al. indicates that up to 130 genes have been identified in humans that are associated with DNA repair. There are a group of general genes involved in signaling and regulating the DNA repair, and those associated with distinct repair mechanisms. Correspondingly, a huge number of proteins are involved in DNA repair. A few examples of DNA repair mechanisms will be outlined; the details are to be consulted with the review by Christmann et al. (2003). The purpose of this short outline is, again, to show how complicated and how many genes/proteins are involved in repairing damaged DNA.

12.3.4.1 Simple Repairs

There are several one-step repairs on some DNA lesions. When guanine is methylated at the O-6 position, it loses the ability to hydrogen bond to cytosine and behaves like adenine, so that G-C is mutated to A-T. This lesion can be repaired in a single-step action by O^6-methylguanine-DNA methyltransferase (MGMT). This enzyme is inducible by genotoxic stress.

Another demethylating agent is AlkB, which removes an alkyl group (of which methyl is one) from 1-methyl adenines or 3-methyl cytosines on single-strand DNA. Similar enzymes ABH2 and ABH3 also repair methylated adenine or cytosine. These are α-ketoglutarate Fe(II)-dependent dioxygenases.

12.3.4.2 Mismatch Repair (MMR)

MMR aims to remove and repair base mismatches (i.e., G/T, G/G, A/C, and C/C) and others caused by spontaneous or induced base deamination, oxidation, methylation, and replication errors. MSH2-MSH6 (MutS homologous protein) is an ATPase, and is conformationally changed upon ATP hydrolysis and binds to DNA. Then, MHS2 binds further with mismatch repair proteins MLH1 (MutL homolog 1) and PMS2 (mismatch repair endonuclease). Exonuclease I then excises the wrong base, and it is followed by DNA polymerase δ and ligase to restore the proper base there.

12.3.4.3 Repair of DSB

Double-strand breaks (DSB) are highly potent inducers of genotoxic effects and most often caused by radiation. A single nonrepaired DSB on an essential gene is sufficient for inducing apoptosis. A special DSB, however, is created intentionally in the meiotic process to allow some crossing over between parental genes.

Two repair mechanisms are known: homologous recombination repair (HRR) and nonhomologous end joining (NHEJ). The mechanism of NHEJ operates mostly in the G_1-phase and is shown in Fig. 12.8. A heterodimer protein Ku70/Ku80 and DNA-PKcs (DNA-dependent protein kinase catalytic subunit) bind to the cleaved ends of DNA. This assembly activates XRCC4 ligase IV. The broken end is then processed by the MRE11-Rad50-NBS1 complex. The activated XRCC4 ligase IV then completes rejoining of the ends. (NBS = Nijmegen breakage syndrome; MRE11 = meiotic recombination 11 homolog; XRCC4 (X-ray repair complementing defective repair in Chinese hamster cells 4). As suggested in the figure, this mechanism may not quite be error-free, as no reference for the sequence is provided.

The second mechanism, HRR, uses as a template the sister double helix, which is partnered with the damaged double strand in the chromosome. The process involves

Fig. 12.8 NHEJ (Fig. 4 in Christmann et al. 2003)

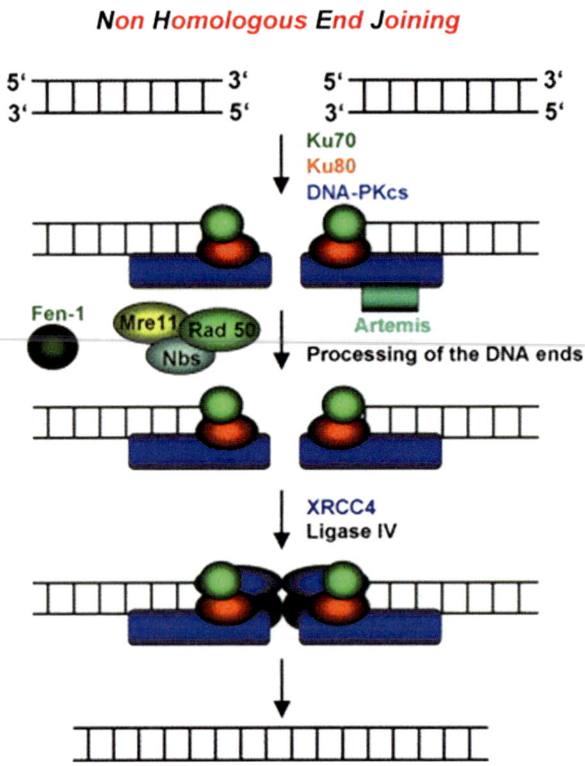

proteins such as MRE11, Rad50/51/52, RPA (replication protein A), XRCC2/3, DNA polymerase, and ligase. ATP-dependent and RPA-stimulated DNA strand exchange takes place, in which a template strand invades base-paired strands of homologous DNA molecules. RAD51 is involved in the search for homology and strand pairing stages of the process.

12.3.4.4 An Overview on DNA Repair

Research on the repair of damaged DNA is making rapid progress. Only a few examples of DNA repair have been mentioned above. It was mentioned at the outset of this section that about 130 proteins were involved in various schemes of repairing damaged DNA. A more recent and detailed review article by Thompson (2012) on the repair of DSB lists about 200 abbreviations for the proteins, enzymes, and others involved at the beginning of the review. That is, even a single type of DNA damage, i.e., DSB, though most serious, requires very complicated schemes using a large number of proteins and others. Other types of damage are possible with DNA; they also require elaborate schemes and numbers of proteins, enzymes, and transcription factors, etc. Radiation could destroy possibly any one or more of the proteins and RNAs involved, in addition to DNA itself, and could hamper the proper repair processes in that way as well. Our understanding of these issues is far from being complete; no, it is just beginning to emerge.

12.3.4.5 Unrepairability of Some DSBs Induced by Radiation

An interesting discovery has been reported about the DSB caused by low-dose radiation (Rothkamm and Löbrich 2003). 1 mGy of γ (or X-)-radiation is considered to be a dose that generates, on average, one track of clustered reactive oxygen species per nucleus, so it was considered to be a minimum dose in order to effect damages. They found a linear relationship between the number of DSBs formed and a dose ranging from 1 mGy to 2 Gy. The DSBs caused by radiation turned out to be more heterogeneous compared to the metabolic DSBs. Some of the radiation-caused DSBs were easy to repair, while others were impossible to repair. DSBs induced in cultures of nondividing primary human fibroblasts by the minimum 1 mGy (X-ray) dose were found to have remained unrepaired for many days, in contrast to efficient DSB repair at higher doses.

The heterogeneity in DSB characters may result from random collateral DNA damage generated near the DSB sites by other free radicals, usually from the same cluster of reactive oxygen species. These and other results about low-dose radiation effects of DNA are reviewed by Bonner (2003).

12.4 Cancer-Causing Processes

12.4.1 Processes Leading to Cancerous States

Cancer is a state of cells that proliferate without restriction. It is caused by the loss of control of cell division. As outlined earlier, many proteins are involved in controlling cell operations. These proteins are under the control of respective genes, and the transcription and protein synthesis mechanisms. Each cell is in contact with other cells and receives chemical signals from the adjacent cells or others, which affects its operation, restricting or enhancing cell division. As outlined earlier, many schemes exist to prevent cell activities that would lead to divisions without repairing damaged DNA. A traditional idea of carcinogenesis is "the accumulation of mutations that lead to the unrestricted cell division". However, other schemes have recently been suggested, including epigenetic effects and cancer stem cell ideas.

Most cancers occur with tissues and organs renewing rapidly. Let's call this type of cancer the "regular type". Other kinds of cancer are also known, with regard to endocrine glands such as the thyroid or the brain, but these are relatively infrequent.

12.4.1.1 Oncogenes and Tumor Suppressors

Some genes produce proteins that deregulate and enhance the cell cycle; these are eventually responsible for the creation of cancerous states. They are called "oncogenes". For example, a single-point mutation in DNA, i.e., a change of G to T, causes a glycine in a normal *ras* product to valine in the expressed protein, and is known to be involved in carcinogenesis. *ras* is an example of a proto-oncogene, and its mutation is an oncogene. Under normal conditions, proto-oncogenes function as a growth factor, receptor, cytoplasmic signaling agent, or transcription factor. Several proto-oncogenes are known in humans. They include: *K ras*, *N ras* (RAS − GTPase), *c myc*, *L myc*, *N myc* (MYC − transcription factor), *bcl-2*, *bcl-1* (BCL = B-cell lymphoma regulator protein), MDM2, *erb-B* (human epidermal growth factor receptor), RET (receptor for tyrosine kinase), NF-2, and PDGF; the lowercase italics represent "gene" and those in uppercase represent "protein". Different genes/proteins are involved in different cancers (p. 373 in Raven and Johnson 2002).

On the other hand, a number of genes are to produce agents to suppress cell proliferation under normal conditions, and their products (proteins) function as cytoplasmic signals, nuclear proteins, or others, as seen in the three options in Fig. 12.7. Damage or mutations on these agents would then allow the cell to proliferate. Tumor suppressors include: APC (adenomatous polyposis coli, Fig. 12.9), DPC4, NF-1, 2 (nuclear factor), MTS1, p53, Rb (retinoblastoma protein), BRCa1 (breast cancer type1 susceptibility protein), etc. (p. 373 in Raven and Johnson, 2002).

12.4.1.2 Accumulation of Mutations: Cancers of Regular Type

A dominant theory of tumorigenesis is that some of the agents (oncogenes, tumor suppressors) which are involved in controlling the cell cycle are to be mutated chemically or otherwise, and sufficient mutations are to be accumulated so that all options, cell cycle arrest, apoptosis, and DNA repair, are to fail, and, as a result, cell proliferation is allowed to proceed without a check. And that is cancer.

This kind of cancer typically occurs with tissues that are rapidly renewing, as in the epithelial lining of the intestinal tract, blood-producing system, male gonad, and others. Starting with a stem cell, it produces progressively more differentiated cells, and the genes will be transmitted along with it. A mutated but not repaired gene would be transmitted to the next stage of the cells. As the renewal of the tissue cells continues, other mutations may be accumulated in the DNA, and, eventually, the normal cell cycle regulations would become dysfunctional, and, thus, a cancerous state begins. An example of colorectal cancer is illustrated in Fig. 12.9. This process may take time, usually a fairly long period, as shown in Fig. 13.3 among the atomic bomb survivors, and is also reflected in the fact that most cancers are diseases among the elderly.

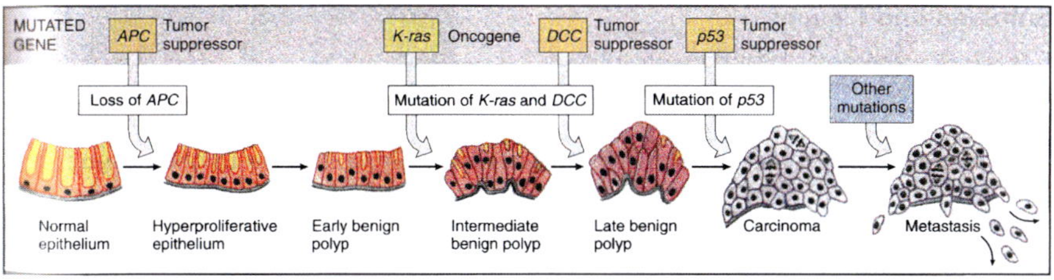

Fig. 12.9 Progression of accumulations of mutations that lead to colorectal cancer (Fig. 18.16, p. 375 in Raven and Johnson 2002)

There is a mechanism involved in stopping tumorigenesis in old age, however. Telomeres are short polynucleotides repeated thousands of times at the end of a DNA molecule. This portion of DNA will not be fully replicated in the division process. As a result, the telomeres will be shortened as a cell is repeatedly divided. After about 30 divisions, many of the telomeres would have been lost, so that no further division becomes possible. Therefore, even if a cell becomes cancerous at this stage, the cell would no longer be able to proliferate. This functions as a brake for carcinogenesis. An enzyme called telomerase, however, restores the lost telomeres and, hence, prolongs the ability of cell division, and may allow carcinogenesis to continue. In normal healthy cells, the expression of this enzyme is suppressed by an inhibitor. Therefore, the damage on this inhibitor could lead to a cancer.

12.4.1.3 Cancers Arising from Fetal Cells as Stem Cells: Cancer Stem Cells

Thyroid and other endocrine and brain tissues are also subject to tumorigenesis, though they are not in the constantly renewing mode as above. Takano (2007) proposed a hypothesis that cancer cells are derived from the remnants of the fetal (progenitor) thyroid cells, instead of normal thyroid follicular cells. That is, a subpopulation (characterized as "fetal" here) of thyroid cancer cells displays properties characteristic of stem cells. This type of cell is termed as "cancer stem cell", and was first discovered in leukemia. Oncogenes play an oncogenic role by preventing the progenitor cells (cancer stem cells) from differentiating. Cancer stem cells are usually not differentiating nor proliferating rapidly, and, hence, are resistant to regular chemotherapy or radiation therapy. However, it keeps its stem cell characteristics, and, hence, contributes to relapse and metastasis (Lin 2011).

12.4.1.4 Epigenesis and Cancer

Epigenesis was outlined earlier; it affects the behavior of cells without changing the sequence of DNA. In normal cells, CG-rich sequences (called the CG island, CGI) in gene promoter regions are generally unmethylated (at the cytosine C-5 position), with the exception of several % CGIs methylated in a tissue-specific manner. On the other hand, the majority of CG sites in repetitive sequences such as ribosomal DNA repeats, satellite repeats, or centromeric repeats are heavily methylated. This limits the access to transcription machinery and stabilizes chromosomes. If this controlled methylation pattern of DNA is disrupted, a tumor may result. For example, an increased methylation (hypermethylation) of DNA in some CGI regions would lead to silencing some tumor suppressor and other genes. On the other hand, loss of methylation at the repetitive sequences has been associated with genomic instability and chromosomal aberrations (Gerhauser 2012).

Another epigenetic effect is the modification of N-terminals of histone proteins. The modification includes acetylation, methylation, phosphorylation, ubiquitinylation, and others, and would change the interactions between the histone core and the DNA wrapping around it. This may contribute to genomic instability, DNA damage response, and cell cycle checkpoint integrity, and, hence, could lead to tumorigenesis. DNA methylation and acetylation of histones are involved in the regulation of CKI (CDK inhibitor). CDK-4 inhibitor p16INK4A is genetically inhibited by point mutations, deletion, or DNA methylation in about 50 % of all human cancers. The hypermethylation of *p16* promoter is found in all the major human malignancies, indicating that p16 is not expressed (Gerhauser 2012).

MicroRNAs (miRNA) are small noncoding RNAs of 20–22 nucleotides, and inhibit gene expression at the posttranscriptional level. They regulate the transformation of mRNA (m = messenger) into proteins, by imperfect base-pairing to the mRNA's untranslated regions to repress protein synthesis or affecting mRNA stability. miRNA have been implicated in cancer initiation and progression (Gerhauser 2012).

12.4.2 Specific Effects by Radiation?

It has been shown that cell operation involves a large number of chemical entities: DNA (in the form of double strand, chromosomes, etc.), mRNA, t-RNA, miRNA, ribosomal RNA, then their product proteins, lipid membranes, carbohydrates, and water. A part of the DNA is transcribed into mRNA, which is translated into a protein. When DNA is damaged, it triggers processes for repair in order to avoid a disaster. Biological evolution has developed such mechanisms, or, rather, species that have managed to develop such mechanisms have survived.

Most biological processes are subject to various disturbances, caused by chemicals, endogenous or exogenous, and biological agents such as viruses. The type and extent of disturbances are governed by the nature of the disturbing factors and the system disturbed. It is highly likely that the biological systems have adapted to the conditions they inhabit. Throughout most of the history of this planet, the surroundings in which life has persisted have consisted of naturally occurring chemicals and organisms, and a very low level of background radiation. How low the radiation was before a century or so ago has not been estimated, but can be

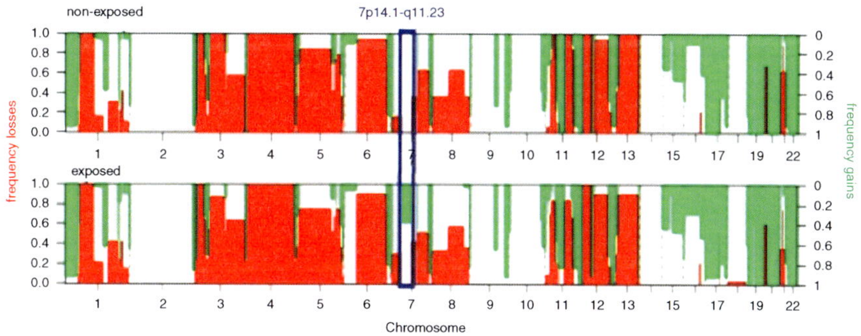

Fig. 12.10 Frequency plot of the copy number changes in the exposed (*lower*) and unexposed cases (*upper*) (Fig. 1 in Heß et al. 2011)

considered to be significantly lower than the present level. And that low radiation has been tolerated by most organisms for the most part of the planet's history, as shown in the next section. This is accomplished by various mechanisms developed through evolution, particularly regarding DNA, which has been outlined in the earlier part of this chapter. However, the probability of a low dose giving adverse effects, strictly speaking, is not zero, no matter how low the dose is.

Now, extra radioactive materials have been introduced into the environment by humans. Would the extra radiation exert any specific effect that the natural chemical or biological agents do not? The answer is very likely "yes", because radiation would attack any molecular species that may include those that cannot be attacked by the chemical and biological agents. Radiation effects have been studied extensively on DNA, and on how damaged DNA would be repaired or otherwise. Radiation could possibly affect all the factors involved in the cell cycle control, including hundreds of proteins: enzymes, transcription factors, oncogenes, tumor suppressors, factors involved in epigenesis, etc., in addition to DNA. As mentioned in Sect. 12.3.4.5, radiation-induced DSBs are somewhat different from metabolically caused DSBs, and quite heterogeneous, and some of them cannot be repaired easily.

Not much study has been conducted on the radiation effects on molecules other than DNA, and they have been discussed in Sect. 12.1.2. No effective means to combat the damage caused by radiation on molecules other than DNA has been found.

An interesting recent discovery is a radiation specific gene aberration (Heß et al. 2011). A main consequence of the Chernobyl accident has been an increase in papillary thyroid carcinoma in the children exposed to the radiation. The researchers examined 52 patients younger than 25 years old who have been exposed to radioiodine (I-131), with comparison to 28 patients of the same age group who have not been exposed. They found that 39 % of the exposed patients had an extra copy of gene 7q11.22-23 on chromosome 7 ($p = 0.0015$). On the other hand, none of the unexposed patients had this extra copy, as shown in Fig. 12.10. The authors suggest that different molecular subgroups and routes of radiation-induced carcinogenesis exist. The CLIP2 gene was specifically overexpressed in the exposed cases. The expressions of several other genes were also found to be correlated with the gain of 7q11.22-23. Hence, the extra copy of this gene may be used as a marker for radiation-induced papillary thyroid cancer in about 40 % of the cases. Obviously, there are, however, other routes leading to the same cancer.

12.5 Are Defense Mechanisms Sufficient?

A crucial question is whether the biological (chemical) mechanisms available in the human body (or the body of any other living organism) can adequately defend against radiation. Acute radiation effects caused by high doses (perhaps >0.25 or 1 Sv) have been acknowledged even by the authorities (IAEA, ICRP, etc.), i.e., the causality is real between the radiation and the effects on biological bodies. It is lethal at high dose (>5 or so Sv). In other words, it has been recognized that the defense mechanisms available in the human body are not sufficient to defend against high-dose exposure, 0.25 Sv or higher.

As argued repeatedly, there is no real defense mechanism against radiation in the sense that such a mechanism nullifies the ionization or other direct effects of radiation on biomolecules. This is impossible. The only thing that living organisms can do and are doing is to repair the damage on DNA, kill and remove the cells with damaged DNA in the process of apoptosis, or provide some chemicals and enzymes to reduce the activities of damaging species, such as hydroxyl free radical.

Most of the disturbing factors on the health of living organisms have been biological and chemical. Defenses against such factors exist, including immune systems. The immune system works against chemical and biological agents only, and not against radiation. It is true, though, that the immune system may help in repairing some damage incurred by radiation, but it is not sufficient to remove radiation effects altogether, as long as the radioactive material remains in the body, because removal of the radioactive material internalized in the body is not within the capacity of the immune system.

Background radiation has persisted throughout biological evolution; such important radioactive factors include potassium (K-39 but includes K-40) and carbon (C-12 but includes C-14) that are present in the natural environment and essential to life, and, hence, incorporated constantly in living bodies. It is likely that most of the living organisms could tolerate this much radiation (see Sect. 11.3.2) caused by K-40 and C-14. How? The regular repair mechanisms developed for other causes may be able to cope with this small amount of radiation.

It is known that K-40 gives about 60 Bq/kg in a normal human body, which amounts to something like 0.3 or so mSv/y. Now, a contentious issue is this: 0.2 mSv/y of Cs-137, for example, is smaller than the dose due to K-40, and, therefore, it is insignificant and can be dismissed as "no problem". This assertion seems to be persuasive enough that many people, including some scientists, believe it. Is it correct?

Potassium is essential to living organisms, and present at something like 2,200 ppm in the human body, and distributed relatively evenly throughout a body, because it is required in all physiology. Therefore, 60 Bq/kg of K-40 is, indeed, distributed more or less at similar levels in all tissues, organs, and cells. (We ignore here the K concentration difference between in and out of cell.) 1 kg of human body contains approximately 10^{12} cells. 60 Bq of K-40 distributes equally in all 10^{12} cells. So, each cell will receive 60×10^{-12} Bq on average. Suppose that 1 Bq emits one radioactive particle per second. Then, over a period of 1 year, each cell would receive 0.002 radioactive particles. Assuming that a single radioactive particle affects about 1,000 molecules (ionize or otherwise), about two molecules in a cell will be affected by K-40 per year. This is a rather small number, and the human body can repair or otherwise cope with this much damage. A similar argument can be made with C-14, which is also universally distributed throughout the body.

Human activities have introduced a number of radioactive nuclides into the environment in recent decades. The major radionuclides artificially introduced are listed in Table 2.1. Cs-137 (and also Cs-134) is one of the most discussed nuclides. Cs is not known to be essential to living organisms, although Cs is usually assumed to behave like K, for they belong to the same group in the periodic chart. If, indeed, this was the case (i.e., Cs behaved just like K), a dose of 0.2 mSv or 60 Bq/kg would not be significant. The fact is that the behavior of Cs in a living system is significantly different from that of K, as shown in Sects. 10.3 and 14.3. Cs would not distribute evenly, but, rather, accumulate in certain tissues and organs. Suppose that 60 Bq of Cs-137 is concentrated in 1 g of a tissue (in 1 kg). In terms of Sv, the effective Sv value is 1,000 times as large. This means 60×10^{-9} Bq per a cell or that 2,000 molecules in a cell of the affected tissue will be bombarded in a year by radiation particles from this Cs-137. This calculation is not meant to be accurate nor precise. No accurate estimation is possible, but this comparison indicates that the radiation effects of Cs (internally present) could be much higher than the nominal Sv value indicates. In other words, it is meaningless to compare 0.3 mSv of K-40 in the human body with 0.3 mSv of Cs-137, as far as their true radiation effects are concerned.

How much radiation will be tolerated by, say, a human body? Well, it is reasonable to assume that living organisms would be adapted squarely to the niche condition. In other words, they would not be equipped with factors more than necessary, because the presence of such extra capability is only burdensome. Human's or other organisms' capacity to correct the adverse effects of radiation would not be an exception. As seen above, the organisms can cope with very small doses (in terms of each cell) of K-40 and C-14. How more extensive this capacity is, twice or three times as large as the minimum, is not known, but it should be quite limited. It might be pointed out, though, that the DNA in a cell operating normally is subjected to disturbing effects of biological and chemical agents more often than those due to K-40 and C-14, and that, usually, it is coping with such disturbances. However, no effective means have been developed in organisms to combat the damaging effects of radiation on biomolecules other than DNA. As a result, diseased states other than leukemia or cancers would appear right after exposure to radiation, particularly in internal exposure, as detailed in Part V.

The internal exposure effects by natural radionuclides (U, Ra, Po, etc.) other than K-40 and C-14 would be similar to the effects of artificially produced radionuclides, because they are not required for life, and, therefore, life has not developed proper ways to handle them. This portion of the background radiation (Table 11.1) would contribute to damaging the health of the organisms.

As often reported, the protracted radiation of a low-level dose is more hazardous than the one-shot radiation of the same total dose (e.g., Cardis et al. 2005 about the nuclear industry workers, and Stewart et al. 1958 about the pelvic X-ray radiation; see the note at the end). This fact implies that the damage caused by a low dose could not be repaired enough before the next low dose and, hence, the damaging effects would accumulate. This is obviously a sign of the limitation of biological defense mechanisms.

It is known that some microorganisms such as *Deinococcus radiodurans* can tolerate high doses of radiation. 10 Gy kills human, 60 Gy kills *Escherichia coli*, but it is said that 5,000 Gy would not kill this bug (*Deinococcus radiodurans*, Wikipedia). This could be due to its higher capacity for the repair of DNA and also the presence of some shielding material against γ-rays. *Homo sapiens* may eventually evolve to tolerate higher radiation doses. In the meantime, many would perish.

Note. The so-called Petkau effect is often cited as a proof for this contention (Petkau effect, Wikipedia). It is highly inadequate, though, to use the Petkau effect as a proof, basically because the system that the Petkau effect is based on is too simple without any delicate systems like DNA and its associated repair mechanisms. It also did not distinguish the behaviors between X-rays and radioactive Na cations.

12.6 Transgenerational Transmission of Radiation Effects?

Would the radiation damage in fathers or mothers be transmitted to their offspring (unexposed), and then their offsprings? Yes, if the damage (mutations in the DNA) by radiation is located in the germ cells. An increased number of malformed babies have been observed among the children of Hiroshima/Nagasaki survivors (Sect. 13.1), and Sect. 14.6 gives examples of transgenerational transmission among the people exposed to the radiation of the Chernobyl incident, particularly the liquidators. The inheritance from fathers must be through the transmission of mutated genes. However, some of the results from mothers may not be genetic transmission, but may, rather, be due to internal exposure to radiation in the embryonic or fetal stages.

How about the subsequent generations? The embryonic cells carry mutated genes transmitted from either the father or the mother or both. If some of the mutations end up in the stem germ cells, oocytes or spermatogonia, without being repaired, those mutations would be transmitted to the next generation of offspring. DSBs are purposefully created in the meiotic process to produce some crossover recombinations between paternal and maternal chromatids. This process is carried out by homologous recombination repair (HRR), and requires cohesin, such as sister chromatid cohesin (SCC) (Lightfoot et al. 2011). Cohesin is a complex protein involved in separating (and binding) sister chromatids during cell division. The lack or mutations of cohesion leads to the failure of DNA damage (mutation) repair. If that happens, the mutations may be transmitted to the next generation.

Abnormal epigenetic states (epimutations) can also lead to tumors and other diseases, as shown earlier. In several cases of colorectal cancer, epimutations have been identified that inhibit the human mismatch repair genes, MLH1 and MSH2 (Sect. 12.3.4.2). Nonmendelain patterns of inheritance suggest that some acquired phenotypes caused by something other than gene mutations, perhaps a pattern of epigenesis, may be inheritable. The epigenetic marks are stable for the adult life of the organism, as the epigenetic pattern can be recreated after somatic cell division. However, epigenetic patterns undergo reprogramming in each generation to restore totipotency to the fertilized egg. And, so far, no strong evidence has been obtained that the abnormal epigenetic states are primary events that occur in the absence of genetic change and are inherited through the generations (Whitelaw and Whitelaw 2008). A possibility has been suggested that small groups of genes (epialleles) escape reprogramming in the gametes and the early embryo. However, no definite proof that this can happen in humans has been obtained.

References

Alberts B, Johnson A, Lewis J, Raff M, Roberts K, Water P (2002) Molecular biology of the cell, 4th edn. Garland Science, New York

Alberts B, Johnson A, Lewis J, Raff M, Roberts K, Water P (2007) Molecular biology of the cell, 5th edn. Garland Science, New York

Bandazhevsky YI (2000) Medical and biological effects of radiocesium incorporated into the human organism. The Institute of Radiation Safety "BELRAD", Minsk

Bonner WM (2003) Low-dose radiation: thresholds, bystander effects, and adaptive responses. Proc Natl Acad Sci USA 100:4973–4975

Cardis E, Vrijheid M, Blettner M et al (2005) Risk of cancer after low doses of ionising radiation: retrospective cohort study in 15 countries. Br Med J 331:77

Christmann K, Tomicic MT, Roos WP, Kaina B (2003) Mechanisms of human DNA repair: an update. Toxicology 193:3–34

Fujimori A, Okayasu R, Ishihara H, Yoshida S, Eguchi-Kasai K, Nojima K, Ebisawa S, Takahashi S (2005) Extremely low dose ionizing radiation up-regulates CXC chemokines in normal human fibroblasts. Cancer Res 65:10159–10163

Gerhauser C (2012) Cancer chemoprevention and nutri-epigenetics: state of the art and future challenge. Top Curr Chem 329:73–132

Heß J, Thomas G, Braselmann H, Bauer V, Bogdanova T, Wienberg J, Zitzelsberger H, Unger K (2011) Gain of chromosome band 7q11 in papillary thyroid carcinomas of young patients is associated with exposure to low-dose irradiation. Proc Natl Acad Sci USA 108:9595–9600

Jorgensen TJ (2007) Cellular defenses against radiation-induced carcinogenesis: cell cycle arrest, DNA repair, apoptosis. NCI REB Radiation Epidemiology Course

Lightfoot J, Testori S, Barroso C, Martinez-Perez E (2011) Loading of meiotic cohesin by SCC-2 is required for early processing of DSBs and for the DNA damage checkpoint. Curr Biol 21:1421–1430

Lin R-Y (2011) Thyroid cancer stem cells. Nat Rev Endocrin 7:609–616

Norbury CJ, Zhivotovsky B (2004) DNA damage-induced apoptosis. Oncogene 23:2797–2808

Ochiai E (1994) Free radical and metalloenzymes: general considerations, Chapter 1. In: Sigel H, Sigel A (eds) Metal ions in biological systems, vol 30. Marcel Dekker, New York

Ochiai E (2008) Bioinorganic chemistry: a survey. Elsevier, Amsterdam

Raven PH, Johnson GB (2002) Biology, 6th edn. McGraw-Hill, Boston

Rothkamm K, Löbrich M (2003) Evidence for a lack of DNA double-strand break repair in human cells exposed to very low x-ray doses. Proc Natl Acad Sci USA 100:5057–5062

Stewart A, Webb J, Hewitt D (1958) A survey of childhood malignancies. Br Med J 1:1495–1508

Takano T (2007) Fetal cell carcinogenesis of the thyroid: theory and practice. Semin Cancer Biol 17:233–240

Thompson LH (2012) Recognition, signaling, and repair of DNA double-strand breaks produced by ionizing radiation in mammalian cells: the molecular choreography. Mutat Res/Rev Mutat Res 751:158–246

Whitelaw NC, Whitelaw E (2008) Transgenerational epigenetic inheritance in health and disease. Curr Opin Genet Dev 18:273–279

Part V
Health Effects of Radiation Revealed So Far

This part reviews the available data regarding the health effects of radiation in several important situations. Since the discovery of radiation by H. Becquerel and M. Curie in the late nineteenth century and of the nuclear fission reaction in 1938, humankind seems to have been fascinated with and hooked on "nuclear" power, ignoring the dark side, i.e., the radiation effects. The serious and very damaging effects on life, human as well as animal and plant, of radiation are emerging in various situations, particularly the Chernobyl nuclear power plant accident (1986). However, the effects in other situations have been rather subtle, requiring careful analyses of the data, and are not very well known. This part tries to outline all the available data. However, exhaustive reports and data are impossible to present in such a small space, so only prominent and representative data are presented.

Chapter 13
Data on Hiroshima and Nagasaki

13.1 Acute External Exposure Effects

The data on the victims of the atomic bombs in Hiroshima and Nagasaki were incomplete, and, besides, had long been treated as classified. First, what doses of radiation received by the people has only been indirectly and inaccurately determined. Figure 11.3 shows only the radiation (γ-rays and neutrons) that would have reached people within about a few kilometers from the hypocenter. This is entirely the instantaneous (at the time of the bomb explosion) radiation, and the effect is strictly external. No internal exposure due to the inhalation of fallout or otherwise has been taken into consideration in the most "authoritative" investigation body (ABCC: Atomic Bomb Casualty Commission, then RERF: Radiation Effect Research Foundation), as will be discussed in the next section.

Unfortunately, the estimations of exposure dose in Hiroshima by the authorities were conducted long after the explosion (August 6), which was followed by black rain (see Sect. 11.5.1) and typhoons (a very strong one on September 17), and were partially based on some experimental data. The radiation must have been significantly reduced because of the circumstances by the time the investigation was carried out, and no serious attention was paid to the fallout effects.

However, there have been reports indicating that many survivors had shown acute radiation symptoms (as illustrated by Dr. S. Hida in Sect. 11.4.4). The Life Span Study (LSS) has followed up 58,500 Hiroshima survivors and 28,132 Nagasaki survivors, and recorded all the relevant details: where exposed, when died, what caused death, etc. The most recent report on this cohort will be discussed in the next section. A few particular sets of data from this report have been analyzed by Shoji Sawada (2011). This study revealed the importance of the fallout effects.

One of the acute symptoms of radiation is the loss of hair (epilation). A set of data concerning epilation among the survivor cohort was analyzed as a function of the external exposure dose, as derived by DS86 by Stram and Mizuno (1989). According to Sawada, the epilation rate (% among the survivors) follows a normal distribution curve up to about 3 Gy (exposure dose), but then somewhat flattens and

DOI 10.1007/978-3-642-38727-2_13,

decreases above 5 Gy. This decrease is believed to be due to the fact that the immediate death effect at the higher level of exposure obscured the epilation incidence. An experimental study was conducted by means of radiation exposure to human head skin transplanted onto immunodeficient mice (Kyoizumi et al. 1998). The epilation rate was found to follow a well-defined normal distribution curve as a function of dose up to 6 Gy, where the epilation rate was 100 %. This result was set as a reference (KSTS) in the analysis by Sawada. Now, the external dose had been obtained as a function of the distance from the hypocenter, as shown in Fig. 11.3, and the rate of epilation among the survivor cohort as a function of the distance has been reported by the LSS (RERF). Sawada deduced the dose from the epilation rate based on the experimental reference data (KSTS) mentioned above, and plotted the dose as a function of the distance, based on the epilation incidence; this is shown in Fig. 13.1. This figure clearly indicates that the exposed dose was, somehow, much higher than the official DS02 (the thin line in Fig. 13.1, which is the same as Fig. 11.3), and was substantial even at a distance of 6 km from the hypocenter, as shown by the solid curve above 1 km. It is believed to represent the exposure to the radiation from the fallout, which was spread far away. For example, the official DS02 data are 0.03 and 0.005 Gy at 2.25 and 2.75 km from the hypocenter, respectively. On the other hand, the estimated doses from this analysis are 1.3 and 1.2 Gy, respectively. That is, the real doses were about 40 and 220 times that of the DS02 initial radiation doses. That is, the official doses were significantly underestimated by ignoring the fallout effects. By the way, this is "external radiation", and "internal radiation dose" is not taken account of, nor can it be reliably estimated.

Sawada analyzed other data sets available on epilation: one set obtained by a University of Tokyo medical team, another by the Joint Commission for the Investigation of the Effects of the Atomic Bomb in Japan (US-Japan), and a set collected by G. O-ho (1957). These data were shown to fit to similar curves as that given by the solid line in Fig. 13.1, suggesting the validity of this analysis.

G. O-ho collected data on acute syndromes right after the bomb in Hiroshima with people who were exposed indoors and did not enter the central region of the epicenter (O-ho 1957). He recorded the incidences of epilation, purpura (severe red rash on the skin), and diarrhea, typical acute syndromes of radiation effects. Sawada analyzed this set of data, and obtained the results shown in Fig. 13.2. This figure shows: (1) that the incidence rate was quite similar with respect to both epilation and purpura, and (2) that the incidence rate of diarrhea was much higher than epilation and purpura. The second point may imply that the first two syndromes are due mostly to external exposure, as presumed, but that diarrhea may have been caused by internal exposure as well, which may have involved not only γ-rays but also β- and α-rays, which were not taken account of in the official estimation of dose (DS02). The plot of estimated doses against distances with regard to these acute syndromes turned out to be very similar to Fig. 13.1. All these results suggest that the fallout radiation effects were, indeed, real and significant, even beyond 2 km, where the initial radiation from the explosion was officially assumed to have not

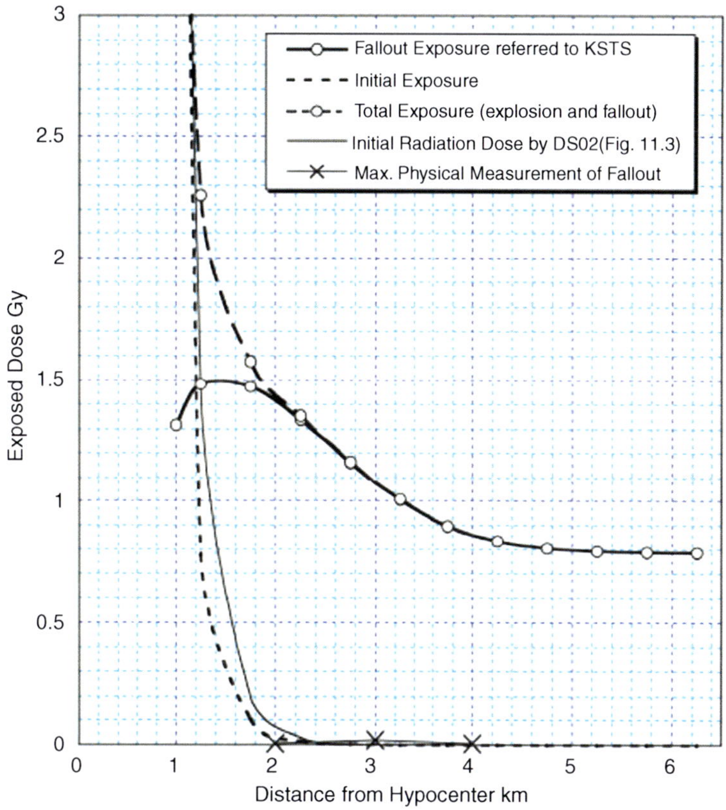

Fig. 13.1 Fallout exposure from the incidence rate of epilation according to the LSS (From Sawada 2011)

reached. These results are very significant in determining the overall radiation effects, because the fallout portion has been ignored in the official account.

Other radiation effects may have been recorded but have not been made public, or no detailed studies were possible in the aftermath of the atomic bombs. Prof. I. Hayashi of Nagasaki University reported that 22.3 % of the babies born from the survivors in 1949–1950 were congenitally malformed, whereas the rate of malformation from nonexposed parents was 8.7 % in the same period. This information is mentioned in a documentary film “The Truth of Fallout” produced in 1957 (H. Kamei director, in Japanese). This film gives details about the detection of Sr-90 (89), Pu-239, Cs-137, and the experimental studies of the radiation effects (death and congenital malformation) on animals and others.

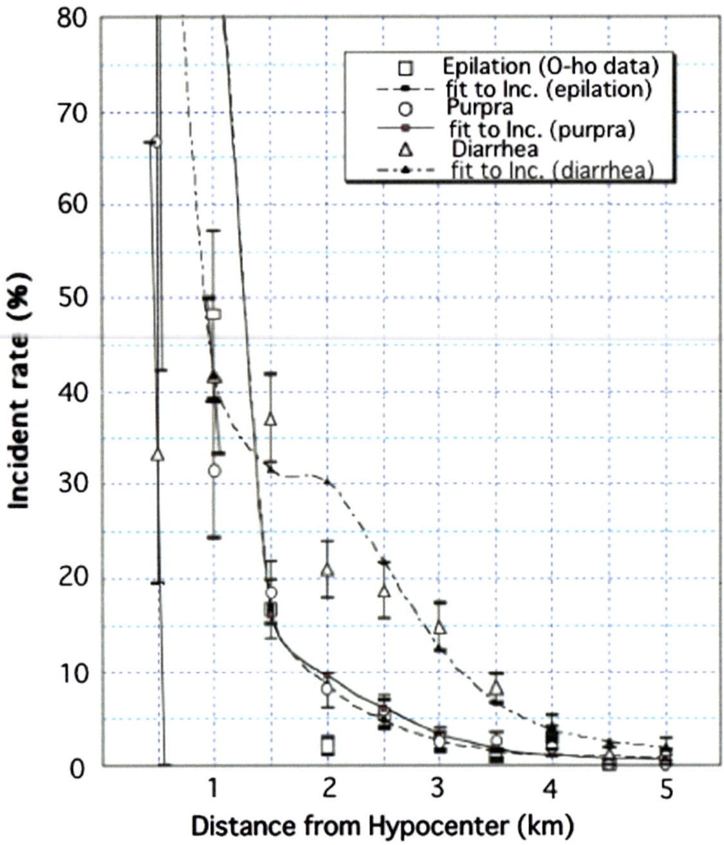

Fig. 13.2 Incidence rates of acute diseases ("Inc." in the figure is a model for the overall dose consisting of the fallout and the initial radiation from the explosion. The parameters involved are determined by fitting a curve to the data; from Sawada 2011)

13.2 Hiroshima/Nagasaki: Long-Term Effects

13.2.1 Slow-Manifesting Cancers: General Trend

Atomic bomb survivors have suffered from various illnesses, particularly various cancers, including leukemia, long after exposure to the radiation. Many of the cancers have taken several decades to appear in some survivors. The cancer rate versus the exposure dose relationship will be discussed in Sect. 13.2.3. Here, a general trend on how slowly various cancers have manifested among the survivors is shown in Fig. 13.3. Leukemia is a special case; it generally appeared around 3–4 years, peaking around 6–7 years after the exposure and then decreased slowly over the next 20 years. Other solid cancers started to increase after about 10 years and then continued increasing as time went by.

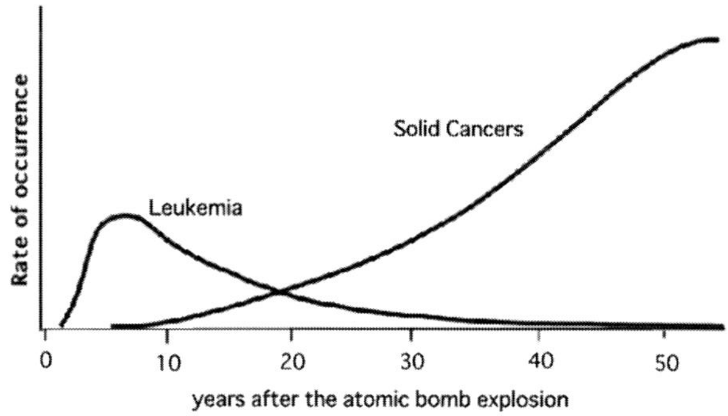

Fig. 13.3 Time sequence of the appearance of leukemia and solid cancers among A-bomb survivors (these curves do not represent the actual numbers, but give conceptual curves)

13.2.2 Radiation-Related Leukemia

Leukemia is the cancer of one of the most vulnerable organs, bone marrow, and has often been found in relation to radiation. Leukemia typically appears 4–5 years after exposure, and then diminishes, as shown in Fig. 13.3. The data of the ABCC were examined carefully by Bizzozero et al. (1966). The rates of leukemia cases, acute, chronic, and probable, were plotted against time (1946–1964) for those exposed within 1.5 km and also for those who were beyond 10 km at the time of the explosion. There are sharp rises at around 1950 in both groups, though the rise in the second group was not as large as that in the first. There was another rise around 1958, which is believed to be related to the large-scale hydrogen bomb tests in the US and the USSR in 1952–1954. In fact, all over Japan, there was a sharp increase of 50 % in leukemia incidence in 1946–1950 as compared to the prewar period, as well as another peak around 1959.

13.2.3 LSS-14 Report

The most recent report on the long-term radiation effects on the cohorts of the atomic bomb survivors of Hiroshima and Nagasaki focuses on the period 1950–2003 (Ozasa et al. 2012 (LSS-14)). The following is a quote from the introduction of this paper.

> The Radiation Effects Research Foundation (RERF), and its predecessor the Atomic Bomb Casualty Commission (ABCC), has conducted a mortality study since 1950 on a fixed population [Life Span Study LSS cohort] of about 120,000 subjects including atomic bomb survivors and residents of Hiroshima and Nagasaki who were not in either city at the time of the bombing to determine the late health effects of ionizing radiation derived from the atomic bombs in Hiroshima and Nagasaki. Periodic analyses of the LSS mortality data have resulted in a series of LSS reports. This is the 14th report in the series, which covers the period of 1950–2003, including an additional 6 years of follow-up since the last comprehensive report. The impact of changing to the DS02 dosimetry system from the earlier DS86 system on radiation risk estimates has been reported for mortality from all solid cancers and leukemia through 2000.

By the way, the dose (DS02) ascribed to each individual in the cohort includes estimated doses for 15 organ sites. These values were obtained incorporating many factors, including the distance from the hypocenter, sex, whether inside or outside of a building, how one was standing or sitting, facing which direction, etc. The accuracy of these estimates cannot be verified, but they have been used as a measure of exposure. It is entirely external exposure, though the dose is ascribed to the dose at an organ, as well as the entire body. This is based on the idea of how an external dose would be distributed among various tissues and organs (ICRP method of "effective" dose estimation of organs). The distribution factor (what % of the external dose would affect a specific organ) is based on the number of deaths due to the failure of a specific organ.

Among the 86,611 subjects with estimated DS02 doses, 50,620 subjects (58 %) died in the follow-up period (1950–2003). 99.6 % of those who were exposed to the atomic bomb radiation aged 40 years or more had died in this period, but almost 80 % of those under 20 years of age at that time were still alive. Twenty-two percent of deaths were due to solid cancer, 1.4 % to lymphoid and hematopoietic malignancies, 71 % to noncancerous diseases, and 5 % to external causes.

The analysis of how vulnerable each person is to the ill effects of the radiation from the bombing is expressed in terms of the excess relative risk (ERR) and is based on the mortality data collected for this cohort. It used a linear model: $\lambda_0(c,s,b,a)$ $(1 + ERR(d,s,e,a))$, where λ_0 is the background mortality number and a = age, b = birth year, c = city (Hiroshima, Nagasaki), d = dose, e = age at exposure, and s = sex. This value represents the number of deaths (for a certain specific cause) and then the ERR may be calculated by fitting this model to the actual mortality. In this analysis, the data on those patients whose exposure dose was very small, less than 0.0005 Sv, were chosen as the background (control group). Under the linear relation assumption, the ERR per dose (Gy) can then be calculated. This can be regarded to represent the slow-developing radiation health risk per Gy.

The ERR/Gy value of all causes of death has been found to be 0.22 (±0.04: 95 % CI), i.e., 22 % higher than the control group, and that for all solid cancers is 0.47 (±0.09). The ERR/Gy values turned to be quite high for cancers of the breast, renal pelvis and ureter, and bladder; 1.60 (95 % CI: 0.99 ~ 2.37), 2.62 (0.47 ~ 7.25), and 1.12 (0.33 ~ 2.26), respectively.

Noncancerous diseases were also affected. The ERR/Gy value is 1.70 (0.96 ~ 2.70) for blood diseases, 0.11 for circulatory and digestive diseases, 0.21 for respiratory diseases, and 0.14 for genitourinary diseases.

A sex difference in radiation effects was also clear. The ERR/Gy value for all causes is 0.15 for males, while it is 0.30 for females. Likewise, the value for all cancers is 0.31 for males and 0.66 for females. That is, females are, in general, about twice as sensitive to radiation as males. The actual number of all solid cancer deaths (EAR = extra absolute risk) in excess of the background number can also be obtained from the collected data. The EAR per 10^4 person-years/Gy is shown in Fig. 13.4. The risk of cancer is higher at lower ages at the time of exposure, meaning that younger people are more sensitive to radiation effects. The figure also shows that the cancer risk increases throughout life.

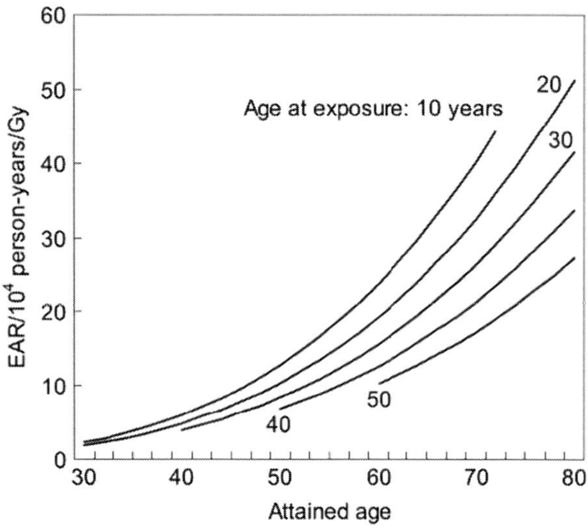

Fig. 13.4 Age dependence of EAR/Gy for solid cancers (From Ozasa et al. 2012)

In the analysis above, linearity is assumed to hold, and, indeed, it turned out to be reasonably so within the analysis, at a relatively high dose range. A crucial question is whether the linearity can be extended to lower doses. This issue has been debated among the researchers in the forms of: (1) whether linearity extends to zero, (2) whether a threshold exists, below which no ill effects would be observed, and (3) whether the ERR versus Gy curve can be a better fit to others than linearity (see Sect. 11.5.6). Many authorities (IAEA, WHO, ICRP, etc.) ascertain that no ill effects will appear below 100 mSv (0.1 Gy). Besides, there are a number of studies that purport to demonstrate the positive (rather than negative) effects at very low dose. This effect is termed "hormesis" (e.g., Feinendegen 2005).

The LSS-14 paper gives a figure of the ERR of solid cancer versus Gy (in the colon), as shown in Fig. 13.5. This figure shows that linearity (L in the figure) is a good approximation at relatively high doses, and seems to be extended to lower doses as well. However, the majority of the points below 0.3 Gy seem to lie above the line. The slope of the L line of Fig. 13.4 is ERR/Gy = 0.4. The value of ERR/Gy deviates from 0.4, and actually increases as the Gy (colon) value decreases below 0.2 Gy, and increases to as high as 1.5 below 0.1 Gy (100 mGy = mSv) (as shown in Fig. 5 in the original paper). This may represent additional radiation effects more than the nominal doses, and are likely the effects of internal exposure doses and "bystander effects". This fact is inconsistent with the idea of "hormesis". The internal exposure is not taken into consideration in the LSS analysis, however.

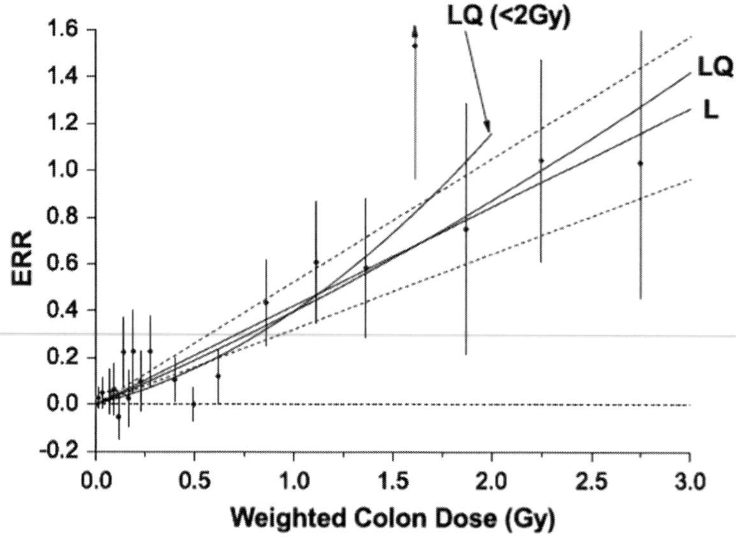

Fig. 13.5 ERR for all solid cancers in relation to radiation exposure dose (From Ozasa et al. 2012)

13.2.4 Problems with the Analysis of LSS Reports

There are essentially two significant flaws in the LSS analysis. One is the choice of the control (background) group, and the other is the neglect of radiation effects other than the primary radiation (neutrons and γ-rays) from the explosion, i.e., neglect of internal exposure.

The choice of control as the survivor group of very low doses (below 0.0005 Gy, farther than 2.7 km from the hypocenter in the Hiroshima case) is very problematic. They were exposed, though at very low doses according to the official valuation. It is highly likely, however, that their exposure doses were higher than the DS02 values, due to fallout and others reasons, as illustrated in Sect. 13.1. If so, their mortality rate would be higher than those who were exposed to no radiation. Hence, the background value in the analysis of the LSS reports is higher than it should be, and the ERR calculated based on this assumption would be lower than the true ERR.

Watanabe et al. (2008) attempted to reevaluate the ERR at very low doses using the data on the unexposed people 0–34 years old in 1945 living in Hiroshima prefecture (the bomb was dropped on Hiroshima city and affected the city and its surrounding areas, but not the whole prefecture) and Okayama prefecture (east and adjacent to Hiroshima prefecture) as the reference (background). They compared their results to LSS-12 (Pierce et al. 1996), not LSS-13 (Preston et al. 2003). They calculated what they call the "SMR" (standardized mortality rate) for the cohort in Hiroshima, separately in terms of the cause of death (cancer), sex, age, and three dose groups; one for very low doses under 0.005 Sv (DS86 used Sv instead of Gy), one for low doses of 0.0005–0.1 Sv, and one for high doses of 0.1–4.0 Sv. The SMR is the ratio of the number of deaths among the exposed population for a certain category to that among the unexposed people of the same category in Hiroshima prefecture or Okayama prefecture.

The SMR value for males for all solid cancer deaths was 1.18 (95 % CI 1.04–1.34) for the Hiroshima cohort of very low dose, 1.20 (1.06–1.35) for those of low dose, and 1.41 (1.20–1.64) for the high dose group. The SMR values for males for leukemia was much higher; 3.15 (1.95–4.87) for the very low dose group, 1.54 (0.82–2.69) for the low dose group, and 3.09 (1.64–5.38) for the high dose group. The corresponding SMR valucs for females turned out to be smaller than those for males, but, except for leukemia for the low and very low dose groups, the SMR values were significantly higher than one. However, it is difficult to make definitive conclusions, because the living conditions including climate and medical care that the people in the background data live under may be quite different from those in the Hiroshima cohort.

In the analysis by Watanabe et al. as well as the LSS studies, the radiation exposure, both external and internal due to the fallout, was not nor cannot be taken into consideration. The external effect of the fallout has been reasonably evaluated by S. Sawada, as discussed in Sect. 13.1. The data on diarrhea, as shown in Fig. 13.2, seem to suggest that internal exposure may be responsible for the much higher rate of the incidence (of diarrhea). However, there are no data available to verify such an internal exposure in the case of Hiroshima and Nagasaki survivors. It is possible, though, that such data can be obtained by examining the bodies of the exposed victims. This kind of investigation has not been conducted. It is highly likely, though, that the cancers found among the atomic bomb survivors long after the exposure have been caused mostly by low-level internal exposure.

13.3 Effects of Fallouts on Other Parts of Japan

Vital statistical data on Japan were collected and analyzed by scientists at Tohoku University, Sendai, Japan (Segi et al. 1965). As seen in Fig. 13.6, many kinds of cancer mortality did increase across Japan after the atomic bombs in Hiroshima and Nagasaki. Pancreatic cancer had leveled off for more than 10 years before 1945, but it shot up to 1,200 % of the pre-1945 level by 1965, though it started to slow down in the early 1970s, after major atmospheric atomic bomb tests had ceased. In other words, this effect may be due not only to the fallout from the Hiroshima and Nagasaki bombs, but also to atomic bomb tests as well as hydrogen bomb tests in the air in the 1950s to the 1970s. In about the same time period, prostate cancer rose by 900 % and lung cancer 750 % compared with the pre-1945 period. The same study showed that the increase in cancers in this period was prominent in two groups: the young (aged 0–19 years) and the old (aged >70 years).

As revealed in Chaps. 14, 15, and 16, the effect of radioactive fallout from atomic bomb tests or otherwise has manifested immediately in an increase of infant

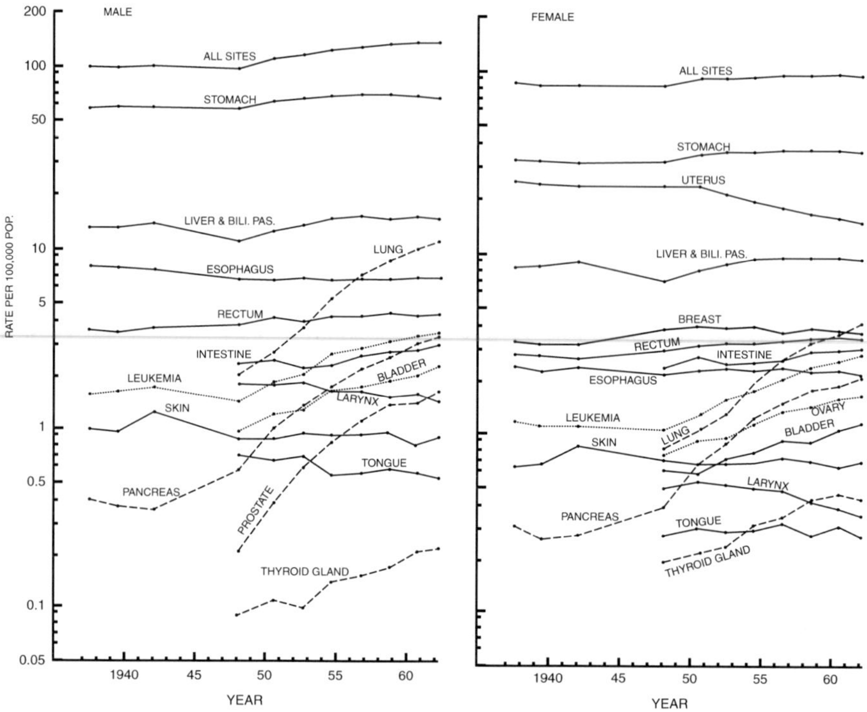

Fig. 13.6 Cancer rates in Japan 1935–1962 (From Segi et al. 1965)

mortality and stillbirths. However, this trend was not seen in Hiroshima and Nagasaki. The data in the previous paragraph suggest that the increase in infant mortality in Hiroshima and Nagasaki was obscured by similar increases in other parts of Japan (Sternglass 1972/1981).

References

Bizzozero OJ Jr, Johnson KG, Ciocco A, Hoshino T, Itoga T, Toyoda S, Kawasaki S (1966) Radiation-related leukemia in Hiroshima and Nagasaki, 1946–1964. I. Distribution, incidence and appearance time. New Engl J Med 274:1095–1101

Feinendegen LE (2005) Evidence for beneficial low level radiation effects and radiation hormesis. Brit J Radiol 78:3–7

Kyoizumi S, Suzuki T, Teraoka S, Seyama T (1998) Radiation sensitivity of human hair follicles in SCID-hu mice. Radiat Res 149:11–18

LSS-14: see Ozasa et al. (2012)

O-ho G (1957) Iji-shinpo (in Japanese) (Medical News), No. 1746, 21–25

Ozasa K, Shimizu Y, Suyama A, Kasagi F, Soda M, Grant EJ, Sakata R, Sugiyama H, Kodama K (2012) Studies of the mortality of atomic bomb survivors, Report 14, 1950–2003: An overview of cancer and noncancer diseases (LSS-14). Rad Res 177:229–243

Preston DL, Shimizu Y, Pierce DA, Suyama A, Mabuchi K (2003) Studies of mortality of atomic bomb survivors. Report 13: solid cancer and noncancer disease mortality: 1950–1997. Radiat Res 160:381–407

Pierce DA, Shimizu Y, Preston DL, Vaeth M, Mabuchi K (1996) Studies of the mortality of atomic bomb survivors. Report 12, Part I. Cancer: 1950–1990, (LSS-12). Radiat Res 146:1–27

Sawada S (2011) Estimation of residual nuclear radiation effects on survivors of Hiroshima atomic bombing, from incidence of acute radiation disease. Social Med Res 29:47–62

Segi M, Kurihara M, Matsuyama T (1965) Cancer mortality in Japan 1899–1962. Department of Public Health, Tohoku University School of Medicine, Sendai

Sternglass E (1972/1981) Secret fallout: low level radiation from Hiroshima to Three Mile Island. McGraw-Hill, New York

Stram DO, Mizuno S (1989) Analysis of the DS86 atomic bomb radiation dosimetry methods using data on severe epilation. Radiat Res 117:93–113

Watanabe T, Miyao M, Honda R, Yamada Y (2008) Hiroshima survivors exposed to very low doses of A-bomb primary radiation showed a high risk for cancers. Environ Health Prev Med 13:264–270

Chapter 14
Data on Chernobyl

The Chernobyl incident took place when the USSR was in power, but during its declining period. It was in a very hard economic situation, a financial crisis. The government was not able to bear the burden of compensating for the damage to the people, the communities, etc. caused by the accident at Chernobyl. Henceforth, they tried to minimize the damage by suppressing the truth of, particularly the health effects due to, the accident. As a result, the data on the damage and health effects during the earlier stages were classified or falsified, or some were apparently even destroyed.

Many scientists and medical doctors were brave enough to gather and analyze data on many aspects of the damage on humans, plants, animals, and the environment. A. V. Yablokov, V. B. Nesterenko, and A. V. Nesterenko collected a huge number of relevant papers and reports, and the New York Academy of Science published their findings as one of their annals (Yablokov et al. 2009). A succinct summary of the health effects of the Chernobyl accident was written by Horishna in 2006 under the auspice of the "Children of Chornobyl Relief and Development Fund (USA)" (Horishna 2006). Independently, German scientists associated with the International Physicians for the Prevention of Nuclear War (IPPNW) have compiled relevant data in "Health Effects of Chernobyl, 25 years after the reactor catastrophe"(Pflugbeil et al. 2011). A Belarusian physician, Dr. Y. I. Bandazhevsky, published a detailed study of the effects of Cs(−137) on health in 2000. The following description is based mostly on these sources.

14.1 Discrepancy Between the Authorities' Reports and the Reality

At the "Chernobyl Forum of the UN" in September 2005, the IAEA and WHO presented an official report on the effects of Chernobyl, which stated that the number of deaths directly related to the accident was less than 50, and that, in the

E. Ochiai, *Hiroshima to Fukushima*, Science Policy Reports,
DOI 10.1007/978-3-642-38727-2_14, © Springer-Verlag Berlin Heidelberg 2014

future, at most, 4,000 surplus fatalities due to cancer and leukemia among the most severely affected groups of people might be expected. In the original WHO report, this figure was 8,930. But the original study from which the WHO quoted indicates a number between 10,000 and 25,000. It is obvious that the official report manipulated its own data, but the authorities have not corrected the discrepancy, not even by 2011. The other authority, the UN Scientific Committee on the Effects of Atomic Radiation (UNSCEAR), declared, in 2011, on the basis of studies carried out during the last 20 years, that the large majority of the population has no reason to fear that serious health risks will arise from the Chernobyl accident. The only exceptions they admit are (1) that thyroid cancer has been prevalent among children due to radioactive iodine and (2) that liquidators have been exposed to high doses of radiation.

The reality of the health effects of the Chernobyl accident is strikingly different from what these authorities present. The details of the reality will be discussed below, but the number of deaths due to the Chernobyl accident alone highlight the discrepancy; the real number seemed to be close to one million by the end of 2004 (Yablokov et al. 2009). Perhaps what the authoritative organizations mean by the number "less than 50" are the deaths caused by the acute effects of radiation on the accident site. This ignores the widespread radiation effects, particularly the internal exposure effects through the inhalation of radioactive aerosol or the consumption of contaminated water, milk, and food.

Another UN organization, the UN Office for the Coordination of Humanitarian Affairs 2000 (OCHA), however, expresses concern with the Chernobyl tragedy, citing more dire pictures of the situation. They also pay serious attention to the long-term health effects. A few lines are cited from the report in the following:

Belarus: Approximately 70 % of the radioactive fallout descended on Belarus, making it the worst contaminated of all the affected countries. Twenty percent of its forests are still contaminated, and the cultivation of 6,000 km^2 of agricultural land has been ruled out by law.

Ukraine: Nearly 3.5 million people, including 1.5 million children, were directly affected by the accident. Half a million children still live in contaminated areas. Nearly 73,000 Ukrainians are now permanent invalids as a result of the Chernobyl accident.

Russian Federation: A total of 200,000 Russians participated in the emergency cleanup operation, of which 46,000 are now invalids.

14.2 Liquidators

Liquidators are the people who were recruited (voluntarily or involuntarily) to deal with the accident, extinguish fire, dispose of and clean up debris, and then construct a sarcophagus to entomb reactor #4. Initially, 200,000 liquidators worked there, and, eventually, the number was increased to about 600,000. They were exposed to high doses of radiation, averaging about 100 mSv; about 10 % of them received *ca.* 250 mSv and a few more than 500 mSv. It is said that 28 died of acute radiation effects within a few months.

The reality is that many liquidators have been suffering from many diseases, including cancers. At the Second World Conference of Radiation Victims in Berlin in 1992, Prof. G. F. Lepin from Minsk (Belarus) stated that 70,000 liquidators were invalids and 13,000 had died. A more recent estimate is that 50,000–100,000 liquidators have died (quoted from Pflugbeil et al. 2011).

The Russian cancer researcher Ivanov (cited in Pflugbeil et al. 2011) found, in a cohort of 47,820 liquidators, that at a medium radiation dose of 128 mGy, the excess relative risk (ERR) was 0.74/Gy in deaths caused by solid malignant tumors, 1.02/Gy in deaths caused by cardiovascular diseases, and 0.42/Gy for all causes. These numbers are comparable to the ERR values found among the atomic bomb survivors (Sect. 13.2.1).

As a result of the increased illness rate, the liquidators who survived following the initial deaths have died, on average, 10–15 years earlier than normally expected. In other words, the ionizing radiation seems to accelerate the aging process. Such an acceleration of aging has been seen especially in the blood vessels in the brain, cataracts, loss of higher intellectual cognitive functions, and loss of stability of the antioxidant system. Mental and psychiatric disturbances are also prevalent among the liquidators exposed to radiation. The details of the health conditions of liquidators after 20 years are summarized in a publication by the PRS/IPPNW Switzerland (2005).

Figure 14.1 depicts the temporal changes of many illnesses from which the liquidators suffered based on the data given in Pflugbeil et al. (2011). The number of cases (per 100,000) of each disease is compared to that in 1986, and it has increased at least 10 times during the period 1986–1993. It has increased 70-fold in the case of diseases of the gastrointestinal tract and 40-fold in the case of diseases of the nervous system and sense organs, endocrine system, and urogenital system. According to Russian information, many invalid liquidators suffer from inflammatory gastroenteritis.

14.3 Cs-137 Internal Effects on Noncancerous Diseases

14.3.1 The Overall Distribution

The problems among the liquidators can be defined somewhat in terms of the exposure dose that has been estimated from the available data. The exposure must have been composed of two factors, external and internal, and the external factor may have been predominant among most of the liquidators.

The health effects on the people in the areas which the fallout descended and contaminated are very likely mostly "internal". The radioactivity of soil of unit area

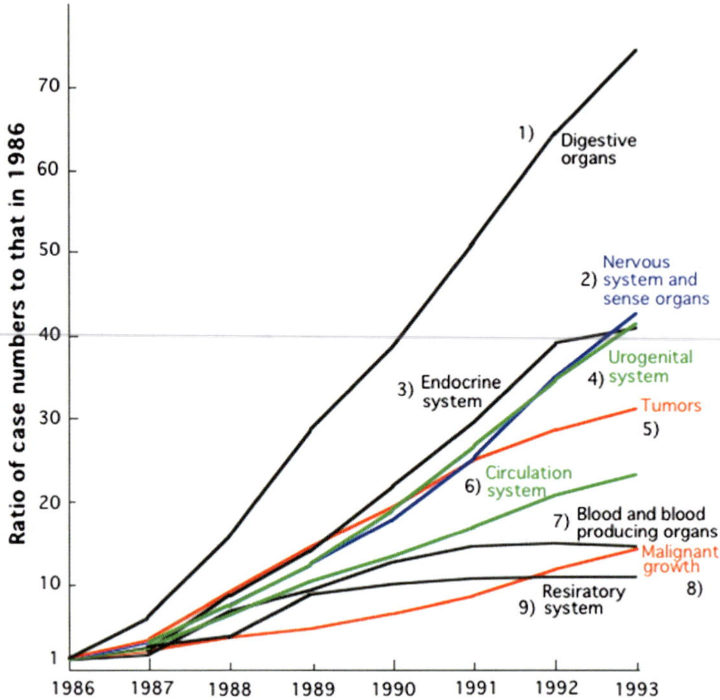

Fig. 14.1 The diseases from which the liquidators suffered (number of cases per 100,000 in 1986: 1: = 82, 2: = 232, 3: = 96, 4: = 34, 5: = 20, 6: = 183, 7: = 15, 8: = 13, 9: = 645; data from Pflugbeil et al. 2009)

can be measured in Bq/m^2. The radioactive material responsible for the radiation may consist of several different isotopes, but, currently, the major contributor has been identified as Cs-137 in the contaminated areas due to the Chernobyl accident. If this radiation impacts on the human body on this ground, the external dose rate has been estimated to be 1.2×10^{-12} Sv/h/(Bq/m^2) by the IAEA/WHO/UNSCEAR (note: a simple conversion would be 3.5×10^{-10} Gy/h(Bq/m^2)).

The degree of contamination, the area of contamination, and the population in the contaminated areas as of 1995 in Belarus, Ukraine, and Russia are summarized in Table 14.1. The fallout from the Chernobyl accident has spread mostly toward the west, northwest, and north. The area in European countries that were contaminated with 37–185 kBq/m^2 are given by Pflgbeil et al. (2011) as: Sweden 12,000 km^2, Finland 11,500 km^2, Austria 8,600 km^2, Norway 5,200 km^2, Bulgaria 4,800 km^2, Switzerland 1,300 km^2, Greece 1,200 km^2, Slovenia 300 km^2, Italy 300 km^2, Moldova 60 km^2.

If the IAEA formula is correct, the midrange contamination 555 kBq/m^2 would amount to 6.7×10^{-7} Sv (Gy)/h. If one lives and is exposed to it 24 h a day, the annual exposure would be 0.6 mSv (mGy)/year. This is relatively low. However, as seen in the next section, the health effects on the people in these contaminated areas

Table 14.1 Population distribution in the contaminated areas

Cs-137 (kBq/m^2)	Belarus	Ukraine	Russia	Total
37–185	1,543,000	1,189,000	1,654,000	4,386,000
185–555	239,000	107,000	234,000	580,000
555–1,480	98,000	300	95,000	193,000
Total (population)	1,800,000	1,296,300	1,983,000	5,159,300
Contaminated area	62,400 km^2	42,000 km^2	57,650 km^2	162, 050 km^2

are much more widespread and serious than this dose number implies. It is very likely due to the fact that radiation effects are mostly caused by internal exposure. In the following description, the author Bandazhevsky uses "Bq/kg" instead of "dose (Gy or Sv)" to indicate the level of exposure, as was argued in Sect. 11.1.2 in this book.

14.3.2 Internal Exposure

How many people have had their health affected by the catastrophe? A detailed set of data is found on pp. 58–260 in Yablokov et al. (2009) regarding noncancerous diseases.

How about the effects of Cs-137, specifically? Have we learned anything about the details of the health effects of internalized Cs-137? Bandazhevsky (2000) of Belarus investigated this in detail. He admits that radioactive Cs had existed in the area long before the Chernobyl accident, likely due to the atomic/hydrogen bomb tests; the accident has added onto it. Cs, being water soluble, has seeped into soil and water systems, picked up by plants, including mushrooms and berries, then entered into cow's milk and meat, which are then consumed by people. Lax regulations on the allowable radioactive levels in food have made incorporation of the radioactive substance from food a common occurrence. The following account is mostly based on Bandazhevsky (2000). It should be noted that no other systematic studies concerning the issues discussed here have been published and that, hence, the results obtained must be viewed with caution, and that further studies are definitely needed.

The internal exposure was measured by using whole-body counters. It measures the γ-radiation coming out of the entire body, and estimates the exposure to radiation in the body (in terms of Bq/kg (of body)). Under the circumstance of constant exposure, the internal Cs seems to increase with age. One set of data showed that the average Cs-137 concentrations as of 1996 was 120 Bq/kg for those born in 1978–1981, 80 Bq/kg for those born in 1982–1988, and 60 Bq/kg for those born in 1989–1996 among residents of the Vetka district, the contamination of which ranges from 185 to 1,489 kBq/m^2. Children's accumulation of Cs-137 in the same district ranged from about 30 to 80 Bq/kg in 2000. Children in other districts were also internally contaminated by Cs-137.

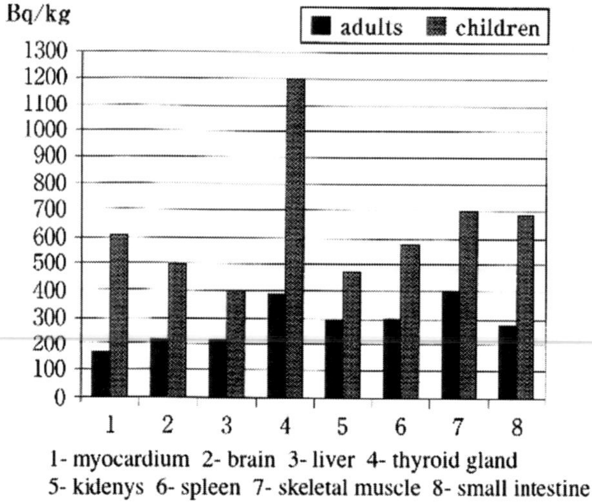

Fig. 14.2 Cs distribution among organs/tissues in dead bodies (From Bandazhevsky 2000)

It turned out that male animals, including humans, accumulate Cs much more readily than females. And it seems that a barrier exists between the placenta and the fetus for the transfer of Cs, and, hence, a newly born baby does not have much Cs, but Cs is then transferred to the newborn through the mother's milk.

The distribution of Cs among different organs and tissues has been studied with experimental animals. An experimental study on albino rats fed with the daily introduction of Cs showed that the order of its accumulation was: kidneys > myocardium > skeletal muscle > spleen > lung > testicle > liver. Radiometric studies of the autopsied bodies of people who died in 1997 in Belarus showed a similar pattern of Cs distribution, which is shown in Fig. 14.2 (Fig. 7 in Bandazhevsky 2000). Several hundred bodies were examined in postmortem autopsies, and the values indicated in Fig. 14.2 are the averages of all observations. In general, the Cs levels among all the organs in children were higher, by two or more times compared to those in adults. The highest accumulation seems to be in the thyroid gland in both adults and children. This is not iodine, which is known to be concentrated in the thyroid, but Cs seems to accumulate there as well. Cs could be a major chemical form in aerosol spewed out from NPP accidents, and it may also be contributing to thyroid abnormalities. The distribution among organs/tissues is: thyroid > skeletal muscle > kidneys ~ small intestine > spleen > brain ~ heart ~ liver. This distribution seems to be consistent with the result of a theoretical study shown in Fig. 10.2, except for thyroid, which was not included in the simulation study. The Cs level in a particular organ correlated with the cause of death well. In other words, e.g., the heart muscle of those who died of heart and vascular diseases contained more Cs-137 than those who died of diseases of the alimentary canal.

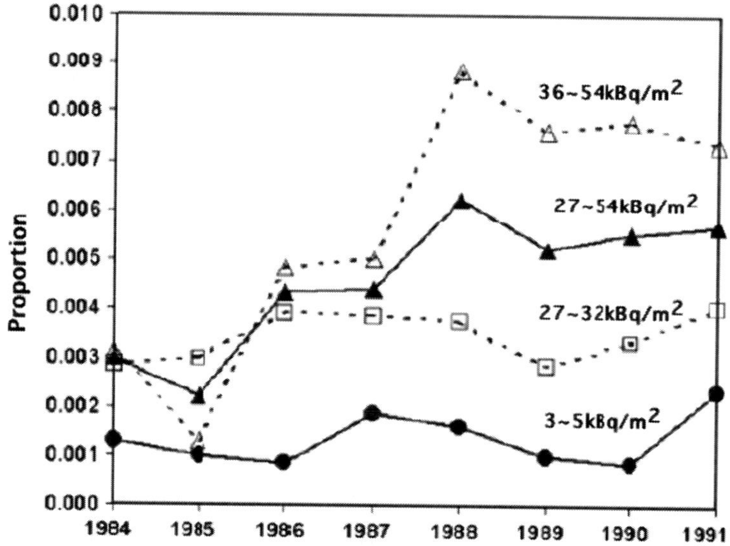

Fig. 14.3 Congenital heart malformation rate in Bavaria (From Scherb and Weigelt 2003)

14.3.2.1 Heart Muscle (Myocardium)

The heart muscle (myocardium) seems to be very sensitive to Cs-137 internal exposure, causing death at 160 Bq/kg in adults. This result was based on over 400 sudden deaths by heart muscle disorders. It has been observed that many children (aged 3–7 years) in Gomel, Belarus, showed cardiac conduction disorders even at lower exposures: about 42 % among children whose internal exposure ranges from 11 to 20 Bq/kg, 48 % for 26–37 Bq/kg, and 73 % for 37–74 Bq/kg.

Several observations are indicative of the nature of the destructive effects of radiation. First, some enzymes in heart muscle cells are reduced in their activities. Alkaline phosphatase and creatine phosphokinase in myocardium cells lost 50–60 % of their activity in experimental animals with 63 $\pm$ 4 Bq/kg of Cs-137. This might be due partly to the destruction of their active sites and mostly to the destruction of some proteins themselves. Cell membranes were also observed to have their structure disrupted, and the integrity of mitochondria was destroyed. Higher Cs levels up to 100–150 Bq/kg has caused yet more serious modifications. Overall, radioactive Cs accumulation in heart muscle results in cell structure damage.

Even in the southern part of Germany (Bavaria), increased rates of congenital heart malformations have been observed, and its rate is dependent on the degree of contamination, i.e., the Bq value of land (Bq/m^2), as shown in Fig. 14.3 (Fig. 8 in Scherb and Weigelt 2003). A study of the postmortem measurement of Cs radioactivity in skeletal muscle was conducted on 300 samples in Graz, Austria, for the period June 1986 to June 1990 (Rabitsch et al. 1991). Ninety percent of these subjects died suddenly from coronary heart diseases. The Cs-137 level in June and

July 1986 was found to be 72.7 Bq/kg (skeletal muscle). However, no measurement was made regarding the radiation levels in the heart.

14.3.2.2 Kidneys, Liver, and Other Organs

The kidneys are where Cs is mainly excreted, like K and Na. It has to go through the glomerulus/Bowman's capsule (filtering) and nephron (reabsorbing/secretion) systems of the kidney, and may be accumulated in the process of elimination because of its less than 100 % efficiency with regard to Cs (refer to Sect. 10.2 and Fig. 10.2). In the dead people of the Gomel district, Cs in kidneys was found at an average of 193 $\pm$ 25 Bq/kg in adults and 645 $\pm$ 135 Bq/kg in children. Microscopic examination revealed a number of pathological damages, especially in nephrons. Necrosis/lysis of the capillary loops in the glomeruli left empty cavities. Other parts in the kidney were also modified by Cs radiation.

A study of albino rats showed that the glomerulus capillaries were invaded by lymphoid phagocytic cells and mesanginal cells proliferated at 63 Bq/kg. These phenomena, as well as the destruction of glomeruli, were exaggerated at higher doses. At 1,000 Bq/kg, structural damage and hemorrhage were prominent.

The livers of the people who died of liver diseases had, on average, 163 $\pm$ 16 Bq/kg (of Cs-137) in adults and 247 $\pm$ 1 Bq/kg in children. Substantial structural changes were observed, as well as disorders in the blood circulation. The Cs-137 concentration in Gomel children exceeding 30 Bq/kg resulted in fewer proteins such as albumin, creatinine, and cholesterin being contained in the blood, indicating disruption of the protein synthetic processes in the liver. Reduced glucose in the blood also indicated disruption of the function of the pancreas.

14.3.2.3 Blood-Producing Systems

Blood is constantly being produced, and, hence, is very vulnerable to radiation. One set of data obtained by Bandazhevsky (2000) showed a correlation between the likely exposure and the red blood cell count. The average red blood cell count in children of age 6–8 years was found to decrease with radiation as follows. The red blood cell count number in children along with the corresponding Cs level (area, Cs level in soil) was: 4.18 $\times$ 10^{12}/L in 106 $\pm$ 31 Bq/kg (Svetilovichi, 555–1,480 kBq/m^2), 4.2 $\times$ 10^{12}/L in 71 $\pm$ 9 Bq/kg (Vetka, 555–1,480 kBq/m^2), 4.27 $\times$ 10^{12}/L in 80 $\pm$ 7 Bq/kg (Stolbun, 185–555 kBq/m^2), 4.40 $\times$ 10^{12}/L in 30.3 $\pm$ 0.7 Bq/kg (Gomel, 40–185 kBq/m^2), and 4.55 $\times$ 10^{12}/L (Grodno, <40 kBq/m^2). It was also observed that the children in the first two of these groups (of Svetilovichi and Vetka) had markedly low numbers of lymphocytes.

14.3.2.4 Immune Systems

Immune systems were also affected. The phagocytosis activity of neutrophils and lymphocytes and IgA (immunoglobulin A) were found to be considerably lower in the children aged 3–6 years in the Gomel region (contaminated area) than those in uncontaminated areas. IgG was about the same and IgM levels were higher. The children with affected immune systems were susceptible to infectious diseases such as tuberculosis (prevalent), viral hepatitis, and acute respiratory diseases (Bandazhevsky 2000).

Based on reviews of 150 reports on this issue, it has been concluded that the depression of thymus function plays a leading role in the pathology of the immune system due to the radiation from the Chernobyl accident (Savyna and Khoptynskaya 1995, cited in Yablokov et al. 2009). The syndrome caused by the suppression of the immune system has been coined "Chernobyl AIDS". Section 5.4 of Chap. 2 of Yablokov et al. (2009) gives a summary of the various studies on immune system diseases. Just a few examples will be cited below.

One hundred and forty-six children and teenagers operated on due to thyroid cancer in Minsk, Belarus, had the following changes in their immune systems: decrease of T lymphocytes (in 58 % of the children and 67 % of the teenagers), decrease of B lymphocytes (42 % and 68 %, respectively), and neutrophilic leukocytosis in 60 % of the children. A survey of 150 Belarus liquidators 10 years after the catastrophe showed a significant decrease in the number of T lymphocytes, T suppressor, and T helper cells.

14.3.2.5 Type I Diabetes

Another prominent phenomenon was an increase of type I diabetes among children in the contaminated area. This disease is due to a lack of insulin, which controls blood sugar levels. It is likely due to the disruption or destruction of beta cells of the pancreas caused by radiation (Cs-137). The incidence rate of type I diabetes in children in a highly contaminated area (Gomel) increased sharply to twice that before the accident (Zalutskaya et al. 2004).

14.4 Cancers

14.4.1 Thyroid Cancer

Thyroid cancer is acknowledged even by the IAEA, ICRP, and WHO to be caused by the radiation effects of mostly I-131 released from nuclear power reactors. Iodine is a specific element required for the thyroid hormone thyroxine, and, hence, is specifically absorbed by the thyroid gland. Cs is most highly accumulated

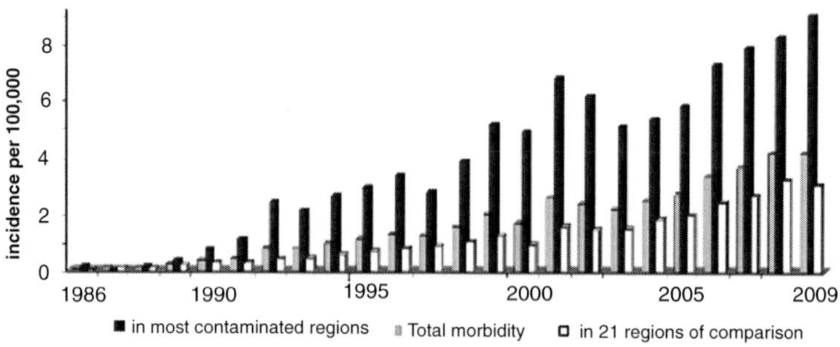

Fig. 14.4 Thyroid cancer incidence rate in the Ukraine (From Ukraine report 2011)

in the thyroid gland in both adults and children, particularly in children, according to Fig. 14.2. Hence, Cs-137 may also contribute to the abnormalities in the thyroid gland. As I-131 has a very short half-life (8 days), it is important for protection (e.g., the administration of KI) to be applied immediately upon its release.

A set of data on the temporal change of thyroid cancers among children 0–14 years of age at the time of the Chernobyl accident in the Ukraine is shown in Fig. 14.4 (from Ukraine report 2011). It shows that a few cases of thyroid cancer seem to have occurred almost immediately within 1 year. In children, the incidence started to increase significantly after about 4–5 years, and has kept increasing, even after 25 years. A similar trend has been observed for the groups aged 15 years or more (Ukraine report 2011). This continuous rise suggests that radiation sources other than the short-lived I-131, such as I-129 and Cs-137, may also be involved.

In a highly contaminated area, Gomel of Belarus, the annual incidence of thyroid cancers among children 2–18 years of age in 1998 was 58 times higher than that in 1973 (before the Chernobyl accident) (Pflugbeil et al. 2011). It needs to be pointed out, though, that cancer is an ultimate result and that radiation more often causes swellings of the glands, autoimmune thyroiditis, and hypothyroidism. These and other anomalies change the hormonal status of children, which may cause defective development in those who are affected.

Thyroid cancer cases among young adults and adults (19–64 years of age) in the Gomel area during the period 1986–1998 was about five times as high as that in the pre-Chernobyl period (1973–1985), indicating that the radiation effects on the thyroid are particularly acute in children (the same increase for children was 58 times, as mentioned earlier).

14.4.2 *Other Cancers*

The extra risk rates for all kinds of cancer were to be compared to those found among the atomic bomb survivors. The exposure doses need to be estimated in order to do this. The whole-body counters were widely used to estimate the dose, though they give only the current Becquerel values. The dose in Gy (or Sv) was then estimated based on the recorded history of an individual's behavior. In any case, the doses have been estimated to range from 0 to 1.5 Gy for the population affected by the Chernobyl incident. It turned out that the cancer risk due to low-level radiation among the affected people was much higher, up 3–10 times, than that found with the atomic bomb survivors (R. Goncharova (2010) cited in Pflugbeil et al. (2011)).

Particularly prominent were cancers of the digestive and respiratory organs, and breast cancer among the women in contaminated areas such as Gomel and Mogilev in Belarus. Overall, 26,000 cases of radiation-induced cancers (including leukemia) were registered in Belarus in the period 1987–1999. The relative (extra) risk values suggested 3–13/Sv, which is an order of magnitude higher than that found with the Hiroshima survivors (Malko 2002). More relevant data on cancers have been collected by Yablokov et al. (Chap. 2, Sect. 6 in Yablokov et al. 2009).

14.5 Other Effects

14.5.1 Infant Mortality

Increase in the infant (prenatal and neonatal) mortality, including stillbirth, miscarriage, and spontaneous abortion, seems to be the most immediate effect of radiation released either from atomic bomb tests or nuclear power reactors. This was recognized by Sternglass, as seen in Chap. 16 from a number of observations in the United States. This is due to radiation, likely from the fallout ingested, affecting the fetus growing in the womb. The fetus, which is growing fast, is the most sensitive to radiation.

The infant mortality increase due to the Chernobyl accident is summarized in Chap. 2, Sect. 7 of Yablokov et al. (2009), and Chap. 3 of Pflugbeil et al. (2011). The following is based on these sources. It has been suggested that the effect of Cs-137 and 134 was immediate, appearing within months after exposure, then leveled off, and that the effect due to Sr-90 appeared only several years later. The first effect by Cs was very likely due to contaminated cow's milk. They did not consider the effects of I-131 and other radioisotopes. This phenomenon, with two peaks in infant mortality after the accident, is exemplified by Fig. 14.5. The third peak around 1994/1995 is said to be due to a change in definition in the vital statistics.

The excess stillbirth rate increased not only in Belarus but also in Ukraine, Russia, as well as some European countries. Its rate in 1987 was 4.3 (/1,000) in Belarus, 2.7 (/1,000) in Sweden, Poland, Hungary, and Greece combined, and 1.2 (/1,000) in Germany. How immediate the effect was is illustrated in Fig. 14.6, which shows the sudden increase of the stillbirth rate in England about 10 months after the Chernobyl accident. The cause of this phenomenon seems to be internal

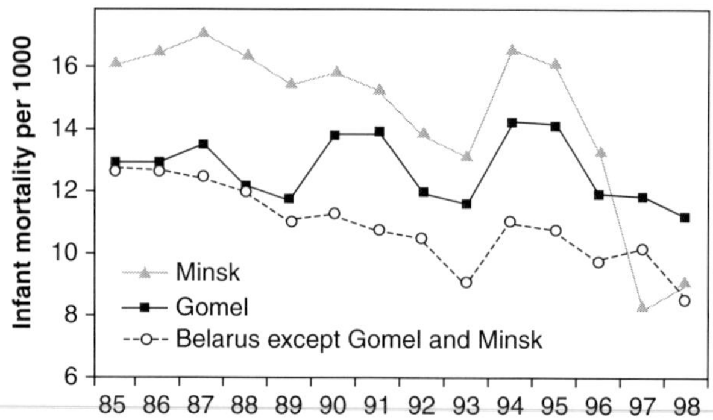

Fig. 14.5 Infant mortality in Belarus (From Fig. 3 in Körblein 2003)

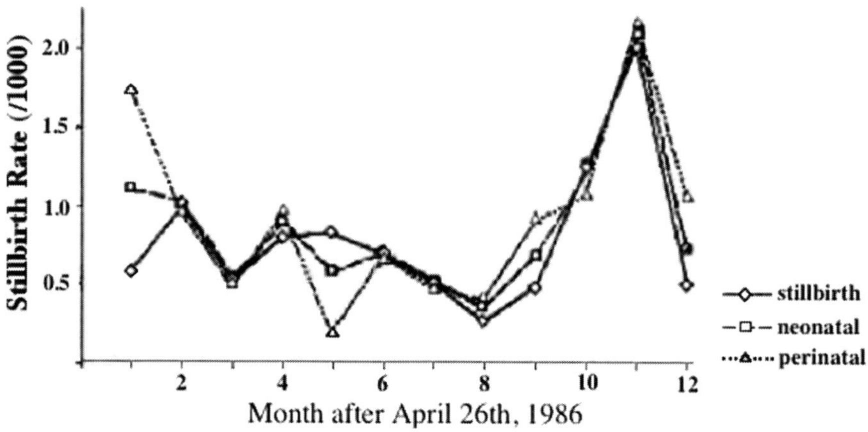

Fig. 14.6 Increase of the stillbirth rate in England (Fig. 7.7 on p. 196 in Yablokov et al. 2009)

radiation on the newly conceived embryo; the radiation could have come from milk contaminated by the fallout.

14.5.2 Teratogenic Effects (Malformations) and Others

The effects on the fetus in the womb will manifest also in congenital malformations of babies following birth. Lazjuk and colleagues (cited in Pflugbeil et al. 2011) published data on the rate of congenital malformations. Before the accident, in 1985, the rate of birth defects in Belarus was 12.5/1,000 live births. This increased to 17.7/1,000 in 1994. This figure is rather deceiving. As the ultrasound test had been developed so that malformations could be recognized before birth, many resorted to abortion. If this number (1,551 cases) was taken into account, the rate for 1994 would have been 22.4/1,000, i.e., it doubled in 10 years. It is reported that

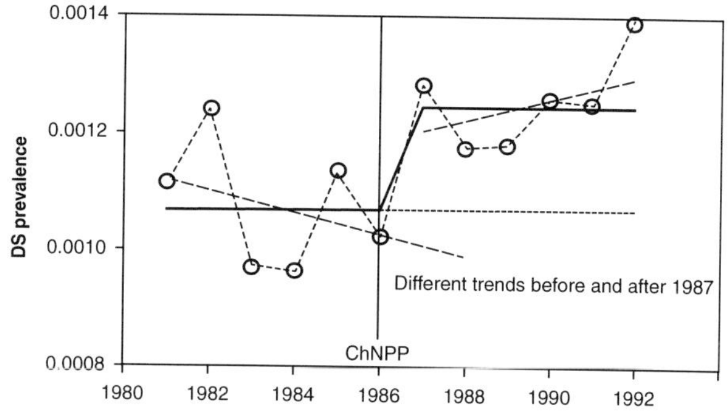

Fig. 14.7 Down syndrome in Europe (From Sperling et al. 2012)

high rates were found with "missing brain", open spine, cleft lip/palate, excess number of digits, and muscular atrophy of the limbs (Hoffman 2001, cited in Pflugbeil et al. 2011).

A study on the populations exposed to chronic low-dose radiation in Polissia, Oblast region, Ukraine (Wertelecki 2010), showed the following results. The overall rate of neural tube defects in Polissia was 27.0/10,000, even for the period of 2000–2006. The rates of conjoined twins and teratomas were also elevated. More data on congenital malformations are given in Sect. 5.12 of Chap. 2 of Yablokov et al. (2009), including some photographs of malformed babies.

One of the genetics-related disorders is Down syndrome or trisomy 21, which means three copies of chromosome 21. Under normal conditions, this syndrome appears at a rate of something like 0.0015 among live births (this is the rate in the United States). The data gathered from Belarus, Bavaria, West Berlin, Sweden, Hungary, northeastern England, and Scotland showed a sudden rise, *ca.* 20 % increase, in the rate of Down syndrome from 1986 to 1987 (Sperling et al. 2012), as seen in Fig. 14.7. The trend before the accident seemed to be declining, but the trend after the accident was increasing, as the dotted lines imply. The authors (Sperling et al. 2012) believed that this is due to the Chernobyl accident.

14.5.3 Effects on the Brain: Psychological and Mental Problems

A study on nerve cells from both humans and animals in the Chernobyl region showed that many psychological and mental disturbances experienced by liquidators and others were, indeed, caused by the damage to the nervous system due to radiation, not the so-called "radiophobia" first invented by the authorities in the USSR (N. Gulaya cited in Pflugbeil et al. 2011). This is contrary to the traditional belief that the brain is not very sensitive to radiation, as mentioned in Sect. 11.4.1. This trend is about the death by brain damage due to high doses of radiation. The brain, however, has been found to be quite susceptible to low-dose radiation, which disrupts the fine structure of the central nervous system and higher mental activities.

Just a few examples will be cited from Sect. 5.8 of Chap. 2 of Yablokov et al. (2009). A survey on pregnant women, maternity patients, newborns, and children in the highly contaminated area (Gomel, Belarus), where the contamination was 185–2,590 kBq/m^2, found that the incidence of prenatal encephalopathy increased after the accident to two to three times that prior to the accident. The same trend has been observed in the contaminated areas in Ukraine and Russia as well. Children exposed to radiation in utero had high indices of mental disability and were more likely to show low intelligence and mental retardation due to the prenatal irradiation.

It has been shown that there was a significant difference in IQ score (verbal) between the children evacuated from the heavily contaminated Pripyat city near Chernobyl and children in a far less contaminated area, Kiev. A similar deterioration of intellectual ability has been observed in Sweden as well. An analysis of 562,637 Swedes born in between 1983 and 1988 revealed that those in utero during the catastrophe had poorer school grades than those born shortly before and after this period. The impairment was heaviest for those exposed to radiation during 8–25 weeks after conception. The degree of damage seemed to correlate with the exposure dose.

A phenomenon now known as "Chernobyl dementia" has become often observed; it includes loss of memory, convulsions, and pulsing headaches caused by the destruction of brain cells in adults. The impact on the nervous system in liquidators is seen in Fig. 14.1. The brain diseases among them included encephalopathy, disorders of the central nervous system, neurocirculatory dystonia, and vegetative vascular dystonia.

14.6 Transgenerational Effects

The liquidators have been exposed to 50–200 mSv. The number of mutations in the genes was found to be particularly high in the children of the liquidators. These children had a seven times greater number of mutations than that found in their siblings who were conceived prior to the service at Chernobyl. This implies that mutations in a father's gonad have been transmitted to the child. As spermatogonia are rapidly dividing, mutations of DNA in the cells initiated by radiation are multiplied rapidly in the process of proliferation, and the mutated chromosome(s) would be incorporated in the embryo. The effects of mutations may or may not manifest immediately, but may show up several generations later (Pflugbeil et al. 2011).

14.7 Effects on Plants and Animals

14.7.1 Plants

The effects of the radiation from the Chernobyl accident on plants and animals are detailed in Sects. 8, 9, 10, and 11 in Chap. 3 of Yablokov et al. (2009). Some

Table 14.2 Radioactivity (Bq/kg) in the leaves of trees

	Ce-144	Cs-137	Cs-134	Ru(Rh)-103	Zr-95	Nb-95
Betula verrucosa	21,000	3,400	1,540	10,290	11,400	18,500
Pinus sylvestris	18,800	4,300	2,100	7,100	6,500	9,900

prominent features will be cited below. Table 14.2 gives data on several radionuclides (Bq/kg of dried weight) contained in the leaves of two tree species in Kiev city at the end of July 1986. In general, more radionuclides accumulated in the root systems (up to sevenfold more than in above-ground parts of plants).

The transfer ratio (TR) is defined as TR = (Bq/kg of plant biomass)/(kBq/m^2 of soil), and is a measure of how readily a radionuclide is absorbed from soil and accumulated in a biological system. TR values have been determined for many plants and many situations. Some TR values for Cs-137 were as follows: wild plants (e.g., *Ledum palustre*) 451, grass (e.g., *Polygonum hydropiper*) 122, fruits (e.g., *Vaccinium myrtillus*) 159, and leaves (e.g., *Fragaria vesca*) 73. TR values for Sr-90 were 14–15 for wild berry (*Fragaria*), 0.6–0.9 for blueberry, and 0.9 for raspberry. TR values of Cs-137 for mushrooms ranged from 1 to more than 100. Many such data have been collected. Obviously, the level of radionuclide accumulation as well as the TR value depend on the soil character, particularly the level of potassium, climate, particular biosphere, season, species, population, etc.

Many morphological changes have been observed with plants. For example, when top buds die (due to radiation), the activity transfers from the apical bud to axial buds. As a result, the newly active axial buds produce extra shoots, leaves, and flowers. This phenomenon has been observed in the shoots of pine, spruce, and others. Tumor-like tissue has also been observed; e.g., such a tissue was seen in 80 % of milk thistle plants growing in heavily contaminated areas. Chromosome aberrations have been observed more frequently in plants growing in the more highly contaminated soil.

14.7.2 Animals and Microorganisms

A. V. Yablokov states as a summary in Sect. 10, Chap. 3 (Yablokov et al. 2009): "chronic low-dose contamination (as the result of Chernobyl) has resulted in morphologic, physiologic and genetic disorders in every animal species that has been studied – mammals, birds, amphibians, fish and invertebrates". It is very likely that animals have suffered from the radiation effects just like human beings, and that the effects would be even more severe than those on humans, as they do not have much of the protection that humans may have.

By September 1986, the population of rodent species in the heavily contaminated Ukraine areas had decreased as much as fivefold. There was an increasing incidence of embryo deaths over 22 generations of bank voles from the contaminated areas. Observations for the period 1978–1999 among

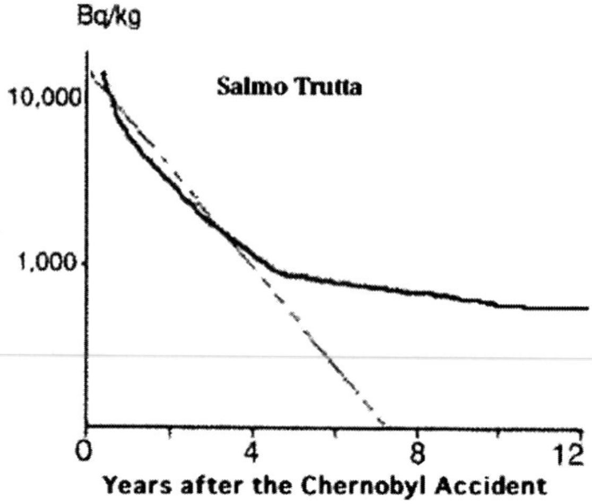

Fig. 14.8 Cs-137 concentration in lake trout in Norway (Fig. 10.4 in Yablokov et al. 2009)

527 horse-breeders indicated that the success of breeding free-range horses correlated with the level of farm radioactive contamination: the greatest number of abortions, stillbirths, and sick foals occurred in a horse-breeding center in the Gomel area, Belarus (where the contamination was up to 1,480 kBq/m^2).

Cs-137 in lake trout (*Salmo trutta*) and charr (*Salvelinus alpinus*) in a northern Norway lake did decline more or less linearly in the first 4 years after 1986, but slowed down unexpectedly after that, as shown in Fig. 14.8 (for trout) (Jonsson et al. 1999 cited in Yablokov et al. 2009).

Some observations on the genetic changes are quite significant. The number of genomic mutations in a population of bank voles increased up to the 12th generation after the catastrophe, despite a decrease in background radioactivity. The offspring of female bank voles captured in the contaminated territories and raised under contamination-free conditions showed the same enhanced level of chromosomal aberrations as the contaminated mothers. The level of both somatic and genomic mutations in a population of barn swallows in the Chernobyl zone was two to ten times greater than in other populations in Ukraine or Italy.

Not much study has been carried out regarding the radiation effects on microbiota. A. V. Yablokov states, in Sect. 11: "Of the few microorganisms that have been studied, all underwent rapid changes in the areas heavily contaminated. Organisms such as *Tuberculosis bacilli*, hepatitis, herpes, tobacco mosaic viruses, cytomegalovirus, soil micromycetes and bacteria were *activated* in various ways.... Compared to humans and other mammals, the profound changes that take place among these small live organisms with rapid reproductive turnover do not bode well for the health and survival of other species." Please refer to Sect. 11 of Yablokov et al. (2009) for more details.

References

Bandazhevsky YI (2000) Medical and biological effects of radiocesium incorporated into the human organism. The Institute of Radiation Safety "BELRAD", Minsk

Horishna OV (2006) Chornobyl's long shadow – health consequences of the Chornobyl nuclear disaster, a summary of findings, update, 2006

Körblein A (2003) Strontium fallout from Chernobyl and perinatal mortality in Ukraine and Belarus. Radiat Biol Radioecol 43:197–202

Malko MV (2002) Chernobyl radiation-induced thyroid cancers in Belarus. In: Imanaka T (ed) Recent research activities about the Chernobyl NPP accident in Belarus, Ukraine and Russia, KURRI-KR-79. Kyoto University, Kyoto, pp 240–256

Pflgbeil S, Claussen A, Schmitz-Feuerhake I, IPPNW, Germany (2011) Health effects of Chernobyl

PRS/IPPNW Switzerland (12 Nov 2005) Health of liquidators (clean-up workers) 20 years after the Chernobyl explosion

Rabitsch H, Feenstra O, Kahr G (1991) Radiocesium levels in humans over a four-year period. J Nucl Med 32:1491–1495

Scherb H, Weigelt E (2003) Congenital malformation and stillbirth in Germany and Europe before and after the Chernobyl nuclear power plant accident. Environ Sci Pollut Res 10(1):117–125

Sperling K, Neitzel H, Scherb H (2012) Evidence for an increase in trisomy 21 (Down syndrome) in Europe after the Chernobyl reactor accident. Genet Epidemiol 36:48–55

Ukraine report: Ministry of Ukraine of Emergencies (2011) Twenty-five years after Chornobyl accident: safety for the future. KIM, Kyiv

Wertelecki W (2010) Malformations in a Chernobyl-impacted region. Pediatrics 125:836–843

Yablokov AV, Nestrenko VB, Nestrenko AV (2009) Chernobyl: consequences of the catastrophe for people and the environment. Ann NY Acad Sci 1181:E1–E39

Zalutskaya A, Mokhort T, Garmaev D, Bornstein SR (2004) Did the Chernobyl incident cause an increase in Type 1 diabetes mellitus incidence in children and adolescents? Diabetologia 47:147–148

Chapter 15
Data on Fukushima

Only 24 months have passed since 2011.03.11 at the time of writing. Therefore, not much data have yet become available regarding the health effects of the Fukushima Dai-ichi NPP accident. The trends seen in the following reports must be viewed with caution, because the sizes of the data sets are often not large enough to allow scientific, i.e., statistically meaningful, analysis.

15.1 Infant Mortality

Körblein (2012) collected data on the infant mortality rate in Japan over the last 10 years. The result is shown in Fig. 15.1. There is a clear declining trend up to the accident date (the vertical broken line indicates 2011.03.31), and then a sudden increase is seen. A replotting of the same data referenced to the declining trend demonstrated that a few points after 3.11 in 2011 were outside of the variations (twice the standard deviation σ), while all the other points in 2002–2010 were within 2σ. The same report indicated that the number of births also declined beyond the trend up to 2010. The decline became significant from December 2011, and was also prominent in Fukushima prefecture. The number of diseased deaths of those who were 1–19 years old in Fukushima in 2011 as compared with that in 2010 is shown in Fig. 15.2. However, the number is too small to see whether the difference is statistically significant.

15.2 Heart Failure and Other Diseases

Another set of data available is the vital statistical data of all death causes reported in all prefectures in Japan, though only up to the end of 2011. One set of such data is the death rate per 100,000 caused by heart diseases (except for high blood pressure), and is shown in Fig. 15.3. It shows a general gradual increase (the purple line in the

E. Ochiai, *Hiroshima to Fukushima*, Science Policy Reports,
DOI 10.1007/978-3-642-38727-2_15, © Springer-Verlag Berlin Heidelberg 2014

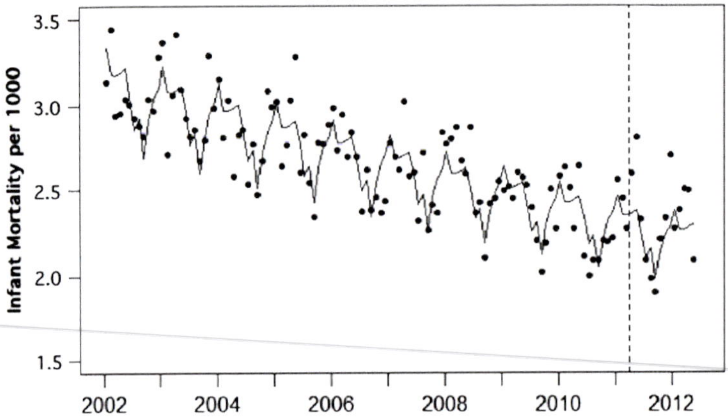

Fig. 15.1 Infant mortality over the last 10 years in Japan (From Körblein 2012)

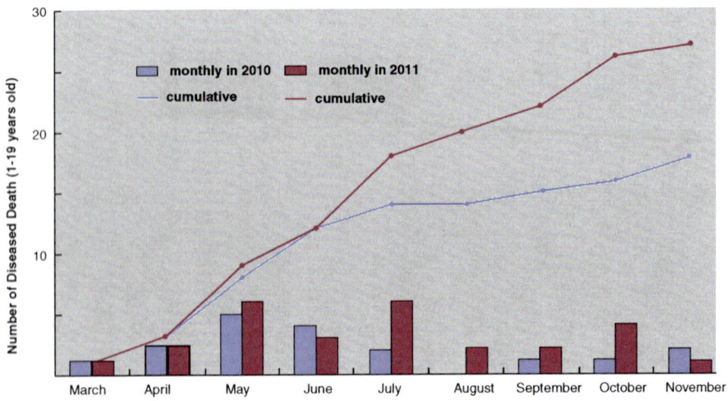

Fig. 15.2 Number of deaths among Fukushima children (1–19 years old) (Produced by S. Nakate (2012), based on the official vital statistic data: http://www.e-stat.go.jp/SG1/estat/NewList.do?tid=000001028897)

figure) in the whole of Japan over the period 1997–2011, and a sudden increase in Fukushima prefecture (red line). A sudden rise, though not so prominent as that in Fukushima, is also seen in Niigata, Miyagi, and Ibaraki prefectures, which are adjacent to Fukushima prefecture. The increase seems to be continuing into the first half of 2012 according to more recent data.

Incidents of sudden cardiac arrest (heart failure) are increasingly being reported, starting in 2011. For example, Ohara hospital in Fukushima city (50 km northwest of the NPP) reports that the number of cardiac arrests and angina was 143 and 266, respectively, in 2010, which increased to 199 and 285, respectively, in 2011, and in the first six months of 2012, the corresponding numbers were 184 and 212, respectively, i.e., 250 % and 160 % increases from 2010 to 2012 (http://blog.goo.ne.jp/chiba20110507/e/e2b3672f66dab60fe9dedad9771871a4). The doctors in

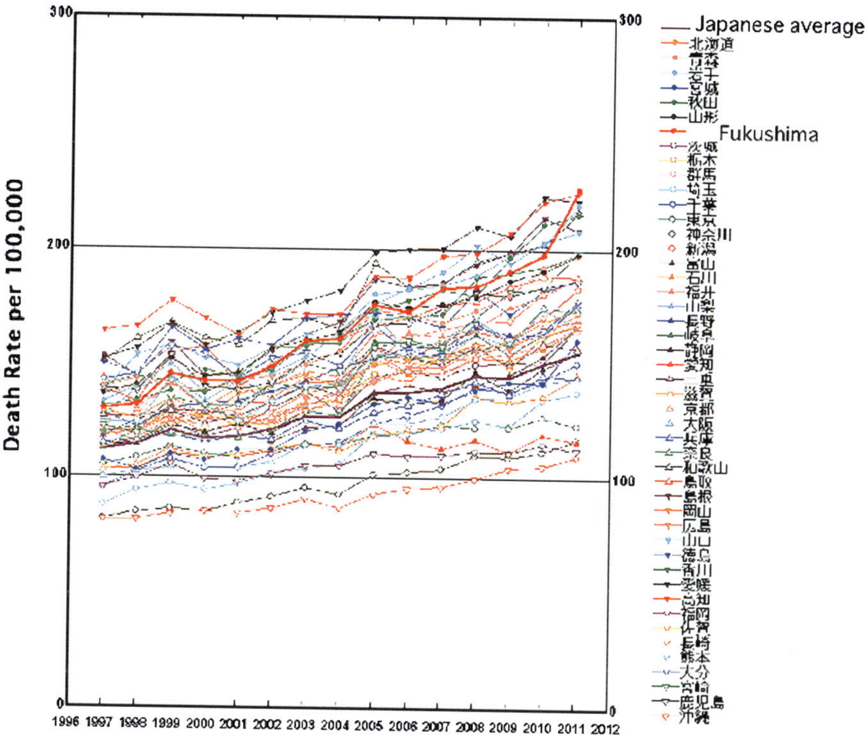

Fig. 15.3 Number of deaths caused by heart diseases in prefectures in Japan; the *purple line* is the Japanese average and the *red line* represents that in Fukushima prefecture, where the TEPCo NPP is located (Created by N. Koshiba, based on the official vital statistics data)

charge suspect that the radiation effects caused these increases. Increases in heart diseases have also been observed in the adjacent prefecture, Miyagi, though the causes have been attributed to stress caused by the disasters (earthquake and tsunami) by the authors (http://www.med.tohoku.ac.jp/d_report/doc2/14-08.pdf).

No official data are yet available, but a large number of cases of adverse health effects have been reported, such as severe nose bleeding, hair loss, severe fatigue syndrome, acute heart failure, acute lymphatic leukemia, systemic collagenosis, spontaneous abortion, and congenital malformation. Another general trend has been noticed by many, although it has not yet been scientifically analyzed. This is the prevalence of viral or bacterial infectious diseases; it might be related to weakening of the immune system in the general public, which has been observed in the Chernobyl incident, as mentioned earlier.

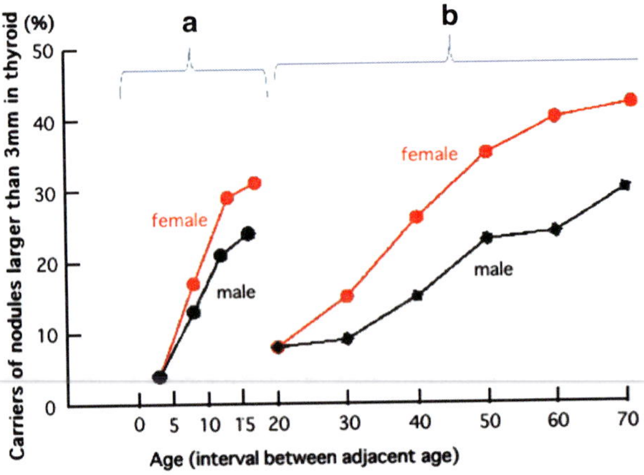

Fig. 15.4 Thyroid anomalies in Fukushima children, as compared to average adults in Japan (Cited from the presentation by M. Matsuzaki (2012)). (**a**) Fukushima prefecture (2011–2012), and (**b**) All adults in Japan

15.3 Thyroid Gland Abnormalities and Cancers

Thyroid abnormalities seem to be widespread among children in Japan. As much as 35 % of 38,144 children less than 18 years old tested in Fukushima were found to carry nodules (<5 mm) or cysts (<20 mm), and 186 of them had larger anomalies in 2011. In 2012, the corresponding numbers increased to 43.1 % and 239 out of 36,000 children tested, respectively (http://fukushimavoice-eng.blogspot.ca/). Recent investigations by various hospitals and organizations have found high numbers of abnormalities in the thyroid glands of children in other parts of Japan as well. How abnormal this situation is can be gleaned from comparable figures among children under normal conditions: the rate of such incidence is 0.8–1.0 % among children younger than 10 years old in the US and other countries. The great difference seen here has been attributed by officials to the following: (1) the definition of abnormalities is different (e.g., sizes of cysts and nodules) and (2) the improvement of instrumentation, i.e., today's instruments can detect smaller abnormalities. Figure 15.4 shows a comparison of the carriers of nodules larger than 3 mm between Fukushima children and the normal adult population in Japan. Too many children seem to carry cysts or nodules in the thyroid gland.

12 thyroid cancer cases in 174000 children younger than 18 years of age in Fukushima prefecture in 2011/2012 have been reported by June 2013, two years after the accident. 15 more are suspected of cancerous states at the time of this rewriting (June 2013; http://www.tokyo-np.co.jp/article/national/news/CK2013060502000133.html). "12 out of 174,000 children" corresponds to 6.9/100,000/ over two years. This number is much higher than that seen in the Chernobyl incident as exemplified by the data in Fig. 14.4. If 15 more likely cases were taken account of, the thyroid cancer incident rate among Fukushima children would be about 7.8/100,000/year, extraordinarily a high rate. (Note: this number is still an underestimate. This number would be 21/100,000/y if the data is more properly analyzed). The authority denies that they were caused by the radiation released from the TEPCo NPP, on the basis that thyroid cancer would emerge only

4–5 years after such an incident. However, the data on the Chernobyl incident show that thyroid cancer did show up even just one year later (see Fig. 14.4), and no probable cause other than radiation effects could be conceived.

15.4 Other Indications

Many indications of radiation effects have been observed with plants, insects, birds, and animals, though not much has yet been studied in terms of systematic ways. One study on a common butterfly *Zizeeria maha* strongly indicates radiation effects (Hiyama et al. 2012). Samples of adult butterflies were collected in May 2011 at ten locations, including several in Fukushima prefecture. Some of the adult butterflies (P-generation) in Fukushima showed relatively mild abnormalities. The first generation of offspring (F_1 generation) showed more severe abnormalities than the original females, and the next generation of offspring (F_2) were worse than F_1. Adult butterflies collected in September 2011 showed even more severe abnormalities. The overall abnormality rate (OAR) of adults caught in May was, on average, 12–13 % in the P-generation, 18 % in F_1, and 34 % in F_2. The OAR values turned out to correlate well with the distance from the Fukushima Dai-ichi NPP and also with the ground-level radiation dose (μSv/h).

Møller et al. (2012) studied the abundance of birds in Fukushima as compared to that in Chernobyl and found negative effects of radiation on the abundance, though the situation in Fukushima was significantly different from Chernobyl.

A symposium was held on March 30, 2013 about the issues of the effects on various living organisms (http://toyokeizai.net/articles/-/13516, in Japanese only). A study on rice plants revealed that plants exposed to the low radiation in Fukushima had anomalies in the DNA-repairing genes and other genes associated with cell protection. More details on the butterflies mentioned above were also reported. A bird (bush warbler) was found to have 5.3×10^5 Bq/kg and a large tumor. Monkeys captured in Fukushima city showed high levels of Cs radiation, which correlated with the level of soil contamination, and they had much lower numbers of lymphocytes than normal.

A compendium was recently published on the implications of the Fukushima disaster on agriculture (Nakanishi and Tanoi 2013). Studies were mostly concerned with the distribution of Cs-137 (and 134) in soil and plants. It appears from the studies included in this book that plant contamination was mostly due to the fallout, which was stuck on the surface, and the contamination was high on the previous plants when the accident took place, i.e., mid-winter. A set of data on rice grain indicated the highest radioactivity in bran. When the grain was milled, the radioactivity was reduced to about 1/10th, and, further, by half via washing. Newly grown plants did not seem to contain much Cs, implying that Cs, once adsorbed into the soil, could not be absorbed easily by the roots. This also seems to be the case with other plants. However, how well Cs in soil is absorbed by roots seems to be dependent on the condition of Cs in soil. It was also found that the fallout contamination was not uniform and quite patchy.

15.5 Current Situations

Sections 8.5.1 and 8.5.2 summarized the release and distribution of radioactive material from the Fukushima Dai-ichi NPP. The melted nuclear fuels have not been retrieved, nor have they been contained and shielded within a secure space. The situation is far more serious; it is virtually impossible for human beings to work in certain areas of the four destroyed reactor buildings because of the high levels of radiation. Radioactive material is still being released into air as well as into subterranean space and rivers and the ocean. The amount of radioactivity released from the NPP was 2.4×10^8 Bq/day as of February 2, 2012, almost a year after the accident (Asahi Newspaper 2012.02.27). More recent data have not been published, nor have data on the amount of radiation released into the ocean been made available. TEPCo has now (summer of 2013) admitted that a large amount of contaminated water has gone to the ocean beside the facility. A major river, Abukuma, running through Fukushima prefecture has been estimated to carry 20 billion Bq of Cs-137 per day into the ocean during the period from June to August 2011.

A sign of contamination of the seawater close to the facility is found in sea creatures living nearby. A recent finding (March 1, 2013) was that some fish contain high levels of Cs-137. No other radioactive isotopes have been examined, but not because they have not been present. A sample of the fish ainame (*Hexagrammos otakii*) contained Cs-137 at 510,000 Bq/kg, and another fish murasoi (*Sebastes pachycephalus*) contained 250,000 Bq/kg. The current Japanese regulation is less than 100 Bq/kg in fish and meat. The Ministry of Environment of the Japanese government reported, on March 14, 2013, that they have found high radiation levels in several wild animal species in Fukushima prefecture.

References

Hiyama A, Nohara C, Kinjo S, Taira W, Gima S, Tanahara A, Otaki JM (2012) The biological impacts of the Fukushima nuclear accident on the pale grass blue butterfly. (Nature) Scientific Reports, 2

Körblein A (2012) Infant mortality in Japan after Fukushima. Strahlentelex mit Elektrosmog Report, No. 622–623, Dec 6

Matsuzaki M http://fukushima-evacuation-e.blogspot.fr/2012/07/a-statement-of-facts-what-is-currently.html

Møller AP, Hagiwara A, Matsui S, Kasahara S, Kawatsu K, Nishiumi I, Suzuki H, Ueda K, Mousseau TA (2012) Abundance of birds in Fukushima as judged from Chernobyl. Environ Pollut 164:36–39

Nakate S (2012) http://blog.goo.ne.jp/chiba20110507/e/f0d4e694489cbb5de91bd16f70ac1390

Nakanishi TM, Tanoi K (ed) (2013) Agricultural implications of the Fukushima nuclear accident. Springer Verlag. http://link.springer.com/book/10.1007/978-4-431-54328-2/page/1

Chapter 16
A-Bomb Tests; Three Mile Island and Other Incidents; NPPs Under Normal Operation; Depleted Uranium Bombs

A few more examples will be given here on the health effects of radiation due to nuclear weapons and nuclear power plants under normal operational conditions, as well as following accidents. Quite a large amount of radioactive material is released even under the so-called normal operating conditions. This is mostly due to the production of gaseous material during the fission process. The gas, mostly radioactive, needs to be vented in order to maintain adequate pressure inside the reactor. For example, in Japan, a nuclear power plant of 1,000 MW is allowed to routinely release 9×10^{14} Bq per year. The health effects of depleted uranium munitions are also serious, and are likely due to radiation from the uranium itself.

16.1 Effects of A-Bomb/H-Bomb Tests

As outlined in Sect. 4.3, a large number of nuclear weapons tests were conducted by nuclear nations. The fallout from those tests is believed to amount to that equivalent to 40,000 Hiroshima atomic bombs. The radioactive material has spread over the entire earth, but especially heavily in the northern hemisphere. Has there been any signs of the health effects of these tests? The answer is that there are many. Only a few examples will be mentioned here. The effects seem to have been the most serious among infants.

The US changed the atomic bomb test site from the South Pacific to Nevada in 1951. There was an immediate devastating effect on the birth weight of newborns. Figure 16.1 shows the data for low birth weight (less than 2,500 g) in Nevada as compared to that of the US as a whole (Gould 1996), showing the sudden increase in 1951. It turned out that the majority in this category (those in the peak in 1951) were white children, who mostly lived in the Las Vegas area. The wind during the tests happened to be directed toward Las Vegas. This experience was recognized and the test dates from then on were chosen so that the fallout would not travel toward the region where the majority of whites lived. As a result, the effect on white children returned to normal but the effect on non-white children continued to be

E. Ochiai, *Hiroshima to Fukushima*, Science Policy Reports,
DOI 10.1007/978-3-642-38727-2_16,

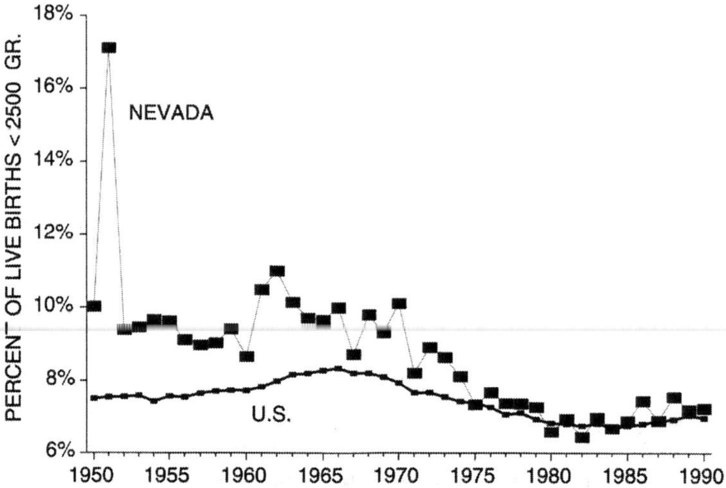

Fig. 16.1 Nevada low birth weight percentage versus US, all races (Fig. 2.9 on p. 43, Gould 1996)

high (Fig. 2.10 on p. 44, Gould 1996). The A-bomb tests were conducted most frequently during the 1960s but continued well into the early 1990s, as shown in Fig. 4.4. The effects of these tests also seemed to be reflected in the rise of the curve of the US in Fig. 16.1.

Warnings based on similar data had been made by R. Lapp, which are described by E. Sternglass (1972/1981), but had been ignored or suppressed by the authorities. One of the stories Sternglass tells in his book is about the city of Troy in New York State. At Rensselaer Polytechnic Institute, Professor H. Clark, in a radiochemistry class, suggested to students to measure the radioactivity around the campus on April 27, 1953. They discovered high readings, 10 to 500 times more than normal, in the area, not only on the campus but in the surrounding townships as well. The professor immediately guessed the source; the debris from the A-bomb test 2 days earlier in Nevada had drifted there and rained down during the heavy rain on the previous day. The follow-up by the Atomic Energy Commission (AEC) concluded, however, that the exposure caused by this incident would have no significant ill effects on health. However, the radioactivity kept increasing thereafter.

Figure 16.2 shows the rates of newborn baby deaths (immediately after birth) during the period 1935–1987 in the US. There was a decreasing trend, as indicated by the smooth curve. The trend was then reversed from around 1951 or so, the year when A-bomb tests commenced in the US. The overall trend seen here is about the same as that seen in Fig. 16.1. In both of these figures, there is a slight sign of increase starting about 1979–1980; this could have been due to the Three Mile Island accident.

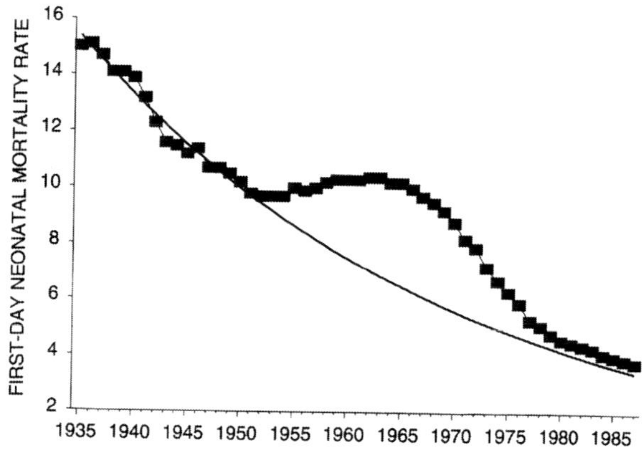

Fig. 16.2 Observed and expected US first-day neonatal mortality rates, 1935–1987 (Fig. 2.8 on p. 42, Gould 1996)

16.2 Three Mile Island Incident

An outline of the accident at the Three Mile Island NPP near Harrisburg, PA, USA, was given in Sect. 8.3. The health effects on the people in the surrounding area have been officially denied. An official stance given on the World Nuclear Association website (WNO; http://www.world-nuclear.org/) is as follows:

> The Pennsylvania Department of Health for 18 years maintained a registry of more than 30,000 people who lived within five miles of Three Mile Island (TMI) at the time of the accident. The state's registry was discontinued in mid 1997, without any evidence of unusual health trends in the area. More than a dozen major, independent health studies of the accident showed no evidence of any abnormal number of cancers around TMI years after the accident....The studies found that the radiation releases during the accident were minimal, well below any levels that have been associated with health effects from radiation exposure. The average radiation dose to people living within 10 miles of the plant was 0.08 mSv, with no more than 1 mSv to any single individual....

Despite the official denial of any health effects, there has been much official (mortality) data that can be interpreted to indicate the health risk associated with the TMI accident (p. 81, Sternglass 1972; Gould and Goldman 1990; Gould 1996). Two sets of such data will be given here. Figure 16.3 shows the changes in mortality from before to after the accident in ten counties around the TMI NPP referenced to the USA average, as compared to those in the entire state of Pennsylvania, where the effects from the accident may be much lower. It clearly shows significant increases in mortality in the counties near TMI in most of the causes of death. It is a matter of debate as to whether this change can be attributable to the accident, i.e., perhaps the radiation released by the accident is not easy to strictly verify. However, can there be any other possible reasons for this change between before and after the accident?

The official documents tend to attribute such changes to people's psyche (fear and stress). Well, people in the immediate vicinity may have felt fear and stress, but

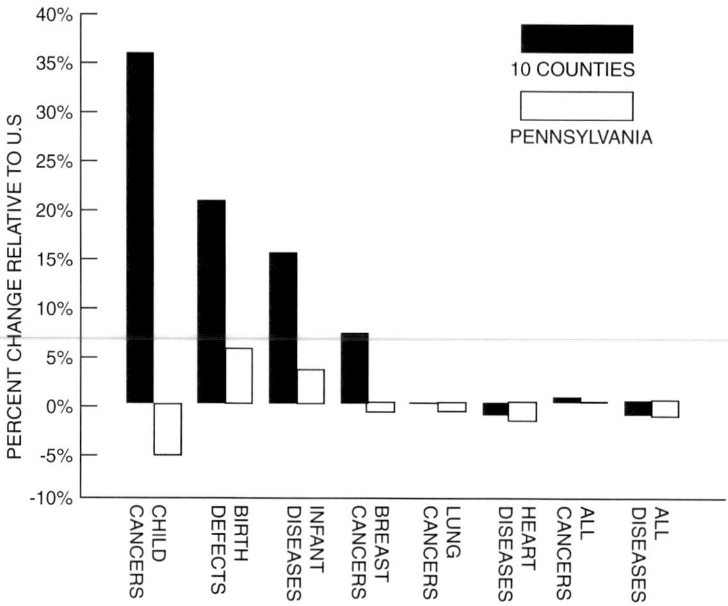

Fig. 16.3 Changes in mortality in ten counties around TMI and in Pennsylvania relative to the US as whole from before the accident (1968–1973) to after (1979–1983) (Fig. 5.5 on p. 68, Gould and Goldman 1990)

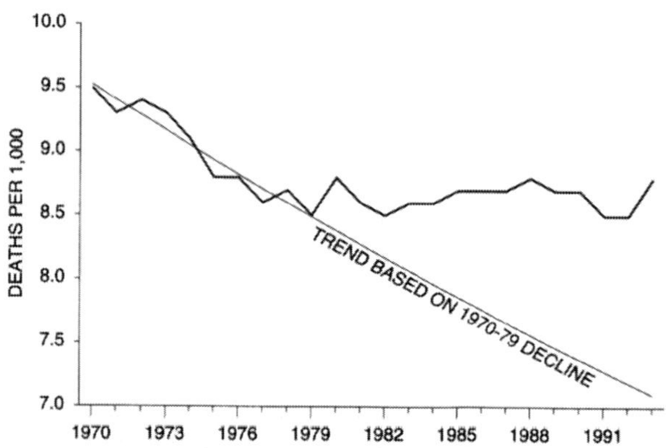

Fig. 16.4 US mortality rates, 1970–1993, observed and expected, 1970–1979 trend (*straight line*) (Fig. 3.13 on p. 84, Gould 1996)

how about the entire population of the United States of America? Did they, even in California or Texas, all have concerns and fears related to the accident? It is highly unlikely. However, the mortality data of the entire population of the US show a sudden increase after the accident in 1979, as shown in Fig. 16.4.

16.3 Effects from NPPs and Other Nuclear Facilities Under "Normal" Operational Conditions

16.3.1 In Europe

In 2002, the German government contracted with the German Childhood Cancer Registry (GCCR) at the University of Mainz to conduct a study of childhood cancers and leukemia in the area around the country's 16 nuclear power plants. They carried out a thorough epidemiological study for the period 1980–2003. The study is known as the KiKK (Kaatsch et al. 2008). The study area included 41 counties in the vicinity of all 16 NPPs in Germany. The following is quoted from a summary report on the KiKK (Nussbaum 2009).

Addresses (distances from the likely spot of radioactive emission (exhaust stack)) of all children younger than 5 years old with leukemia or other malignancies at the time of diagnosis (1,592 cases) were compared to those of three times as many randomly selected children of the same age and sex, residing in the same region who did not have either of these diseases (4,735 controls). The only parameter is the "distance". A logarithmic regression analysis showed a rise of all cancer cases with the decrease in distance, with the sharpest rise within 5 km. It turned out that leukemia risk of the children younger than 5 years old and living within 5 km from NPP was twice or higher than those living farther than 5 km. The increase in leukemia risk remained significant in the <10 km zone compared to the >10 km zone.

What is the reason for this? The authors of the KiKK report state: "In view of the fact that this result was not expected under current radiation epidemiological knowledge, and considering that there is no evidence of relevant accidents, and that possible confounders could not be identified, the observed distance trend remains unexplained."

An external epidemiological review panel on this report pointed out that another sophisticated analysis of the incidence and mortality rates of childhood leukemia near 136 nuclear facilities in the UK, Canada, the US, Germany, Japan, and Spain showed statistically significant increases between 14 and 21 % of leukemia incidence in children younger than 9 years old near many of these sites (Baker and Hoel 2007). The panel concluded that the KiKK study suggests a causal relationship between radioactive emissions from nuclear power plants and the undisputable positive trend of childhood malignance incidence with decreasing residential distance from these plants within a radius of 10 km, and that "there exists no plausible alternative hypothesis" (Joel et al. 2007). A more recent study on French nuclear facilities for the period 2002–2007 (Sermage-Faure et al. 2012) is consistent with these findings: there was a higher rate of leukemia among children living near nuclear facilities.

There have been reports that did not confirm the contentions above. They did not contradict the findings, but suggest that the statistical analysis was not definitive. One of the reasons to reject the causality between the increased leukemia rate and proximity to nuclear facilities is that the radiation doses near the facilities were not measured and that some estimates of the dose were too low for the effects, according to the critics (see Müller). Many of these contentions are based on the current authoritative dictums of the IAEA, WHO, and ICRP, which are essentially based on the Hiroshima/Nagasaki data, which neglect "internal exposure", as discussed earlier.

16.3.2 In the United States

Gould and coworkers extensively studied the breast cancer risk around all the nuclear facilities in the US (Gould 1996). Figure 16.5 shows the age-adjusted breast cancer mortality rates since 1950–1954 in counties within 100 miles of the Oak Ridge nuclear facility in Tennessee, which was the first nuclear facility for the Manhattan Project and continuously in operation. The shaded counties are those downwind closest to the Oak Ridge facility, where the combined age-adjusted rate rose by 38 % from 1950–1954 to 1989. On the other hand, that for the entire US was a modest 1 % for the same period. On the contrary, the eight upwind counties within 50 miles experienced a decline of 4 %, which is not statistically significant. The decline could be within chance occurrence. However, the 38 % rise would occur by chance in only one in a million cases. This figure clearly shows the effect of the prevailing wind on carrying the radioactive emission gas or particles. Similar statistically significant changes in the breast cancer mortality rate since the establishment of the facility have been observed and recorded for all the nuclear facilities in the US in the book by Gould (1996).

Chapter 14 of Sternglass (1972/1981) gives some examples of the effects on infant mortality due to the release of radioactive gases from nuclear power plants under normal operational conditions. The Hanford reactor (Fig. 4.1) that produced plutonium for the Nagasaki bomb started to operate in 1944. The infant mortality in Benton county, where the Hanford facility is located, increased by 160 % from 1943 to 1945. The Dresden reactor in Illinois is of the BWR type and started to operate in 1959. It released 2.6×10^{15} Bq of radioactive material in 1964 and 2.3×10^{16} Bq in 1965, though no accident took place. The infant mortality rate in the county where this reactor was located increased sharply by 140 % from 1964 to 1965. The infant mortality rate in the surrounding downwind counties increased by 48 %, whereas that in the upwind counties declined by 2 %.

Even a small experimental reactor seemed to have released enough radiation around it. TRIGA, an experimental reactor at the University of Illinois, started to operate in 1962, and the infant mortality rate in the city of Urbana where the university is located increased sharply by 300 % by 1965. The death rate from congenital malformations increased even higher, by 600 % in the same time period. Another TRIGA experimental reactor at Pennsylvania State University caused a similar increase in the infant mortality in State College, PA.

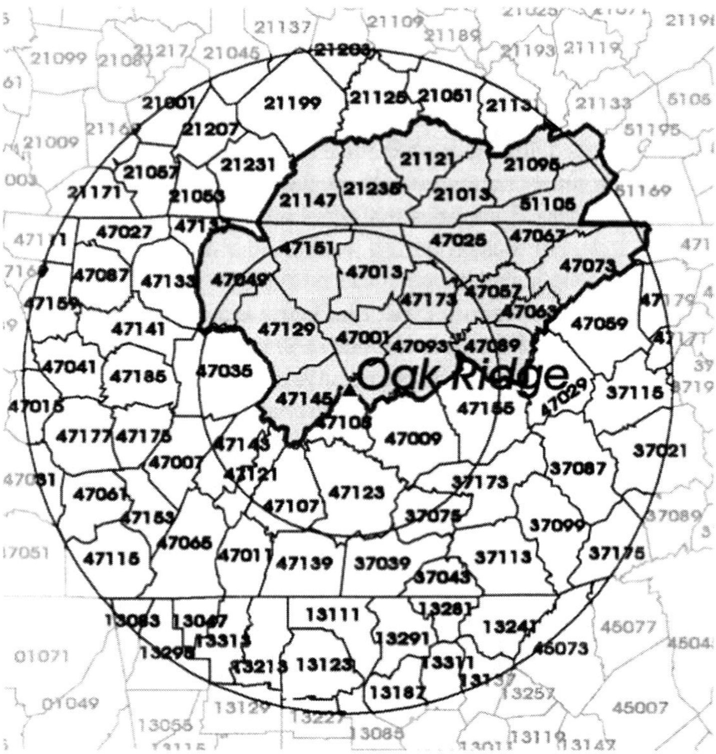

Fig. 16.5 The change in breast cancer mortality from 1950–1954 to 1998 in counties in Tennessee surrounding the Oak Ridge nuclear facility (Fig. 4.1 on p. 90, Gould1996)

The problem of leakage of radiation from nuclear facilities under normal operations has not diminished in recent decades. As a matter of fact, it has increased as the facilities have grown old. A few more recent examples will be mentioned below.

A year-long investigation by Associated Press (AP) uncovered that tritium (H-3) has leaked from at least 48 of 65 sites (source: NRC records). It occurred mostly through corroded pipes, and at a few sites, including two sites in Illinois, contaminated the drinking wells of nearby homes (Huffington Post 2011.06.21). The official records indicated, however, that the level of contamination was not above the EPA regulation level (20,000 picocuries/L = 740 Bq/L). Accidental leakage is not a rare occurrence. Tritium-laden water with 2.6×10^4 Bq/L was leaked from LaSalle, Illinois, in July 2010, and the water that leaked from the Quad Cities facility in Illinois contained 2.8×10^5 Bq/L of tritium (Huffington Post 2011.06.21). Yet, the NRC did not regulate this.

J. R. Sauer (2011) collected and analyzed vital statistical data in the surrounding areas of the large NPPs at Braidwood and Dresden in Illinois. According to his analysis, within 15 miles of the Braidwood and Dresden NPPs, the risk of cancers increased significantly between the 10-year period 1997–2006 and the previous

10-year period 1987–1996; 15.3 % for breast cancer, 18 % for prostate cancer, 34.8 % for urinary cancer, 30.8 % for leukemia, and 30.2 % for brain cancer. The overall cancer rate increased by 17.8 % in these areas, compared to an increase of 9 % over the whole of Illinois. Cancers in children increased by an overall amount of 38 % during the same period.

It has been noted that thyroid cancer incidence is increasing more rapidly than any other types of malignancy, rising about threefold from 4.33 to 11.03 per 100,000 from 1980 to 2006 in the US (Mangano 2009). Similar trends have also been observed in many other developed countries, and the causes for them have been debated. Mangano (2009) examined the data for 46 states (including the Washington district) representing 276 million Americans, and analyzed 500 county-specific data sets involving 202 million people. In the period 2001–2005, thyroid cancer rates per 100,000 ranged from the lowest of 5.4 in Arkansas to the highest of 12.8 in Pennsylvania; the value for Pennsylvania is 44 % above the national average. States which are amongst the highest are Massachusetts, New Jersey, Connecticut, and Rhode Island, besides Pennsylvania, all of which are in the northeast, although New Mexico, Utah, and Montana are among the highest as well. In terms of county, the highest was 21.4 in Lehigh County, PA, 18.8 in Northampton, PA, 18.3 in Rockland, NY, and other counties in eastern PA and adjacent counties in NY. This is to be contrasted with the fact that Pennsylvania's thyroid cancer rate in the mid-1980s was 40 % below the national average. There are nine nuclear reactors in this area, the highest concentration of reactors in the US.

16.4 DU (Depleted Uranium) Munitions

16.4.1 What Are DU Munitions and Where Have They Been Used?

Natural uranium consists of 99.27 % U-238, 0.71 % U-235, and 0.0053 % U-234. All uranium isotopes are fissile, but U-235 is the only one that is usable for explosive nuclear weapons and nuclear power generation. Natural uranium cannot be used for the most commonly used nuclear reactors (BWR, PWR), since the U-235 content is too low. A certain type of nuclear reactor can use natural uranium though. Therefore, U-235 has to be enriched to about 3–5 % in order for it to be used as the fuel for a nuclear power reactor, and more than 20 % for use in nuclear weapons. For example, 11.8 kg of natural uranium is used to produce 1 kg of 5 % enriched (in U-235) uranium, and the remaining 10.8 kg is the depleted uranium, which still contains about 0.3 % U-235. So, ten times as much DU will be produced in the enrichment process. For the purpose of weapons-grade enrichment, a lot more DU will be produced. DU is also produced by recycling the spent nuclear fuel. Thus, an enormous amount of DU has been produced in the nuclear industry. In recent times, it seems that even natural (not depleted) uranium is used to produce

DU weapons (D. Williams). No matter what, DU is essentially an α-emitter with a long half-life.

When DU is made into a metal, it is shiny, hard, and very dense. Its density is 19.1 g/cm^3, much denser (and harder) than lead. Because of these characteristics, it is very effective as a head of munition, which can easily penetrate armored tanks and underground bunkers. Besides, it is cheap, since there is no use for DU and it is plentiful. In the purely metallic form, it is relatively harmless, though it emits α-particles at a very low rate (because of the long half-life). That is why it has been made into a weapon, DU munitions, and the manufacturers considered it safe to handle. DU munitions have been used in the Gulf War (1991), the Bosnia/ Herzegovina war on Serbia, and the Afghanistan and Iraqi war (2001~). DU has also been used as a counterweight in the propeller of a helicopter, for example.

Unfortunately, it can readily be oxidized by air and turns into powdery oxides (UO_2, UO_3 (yellow), U_3O_8, etc.). When it hits a hard surface and penetrates a tank, the bullet would explode and burn the uranium metal and form minute uranium oxide particles that would be spread around. It contaminates the air and water and the environment with α-emitting radioactive material, which may eventually get into human bodies through inhalation and drinking water and other ways.

Several surface soil samples were collected from Bagdad, Basra, and Suweirah (farming area), and uranium isotopes were carefully determined (Durakovic et al. 2004). The level of uranium in some of the soil samples were quite high; the uranium content (U mg/kg soil) was 0.09 in some parts of Bagdad, as high as 35 at Bagdad gate, 76 at Basra, and very high at Suweirah, at a level of 1,030. The observed U-238/U-235 ratios imply that the contamination was caused by DU.

Uranium is chemically toxic as well as radioactive. A review of its chemical toxicity was published in 2004 (Craft et al. 2004). It must be pointed out that, as long as a uranium entity remains in the body exerting its chemical toxicity, it is simultaneously emitting radiation. Therefore, some of the effects attributed to chemical toxicity in such studies might be attributable to the radiation effect or a mixture, i.e., the synergistic effect.

Was uranium spread in the environment as a result of the use of DU munitions? Dai Williams cites the data obtained by Busby and Morgan by collecting and monitoring airborne uranium aerosol at several UK atomic weapons establishments during the period 2001–2003. The results are shown in Fig. 16.6. Peaks of uranium radiation are clearly observed to have occurred following the military activities in the Middle East.

In 2005, workers clearing a decades-old Hawaii firing range found DU fins from the training grounds. The training had been a highly classified program, and the Hawaiian public were not informed. Torishima, a small island off the Okinawa main island (southernmost major island of Japan), was used for the firing test range by the US military and the test included DU munitions as well. The results were devastating, as the island is devoid of any living beings.

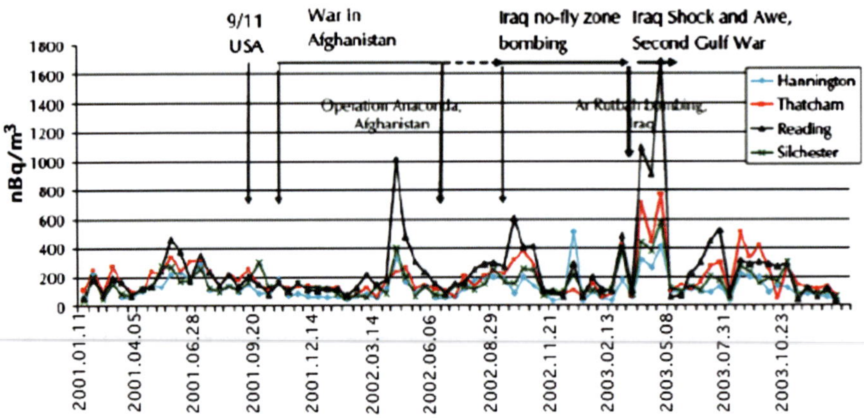

Fig. 16.6 Uranium monitoring results in the UK (Cited in D. Williams)

16.4.2 Gulf War Syndrome Among Veterans

Operation Desert Storm (Gulf War, August, 1990–February, 1991) (Wikipedia: http://en.wikipedia.org/wiki/Gulf_War#Effects_of_depleted_uranium) is reported to have been the first war in which DU weapons, mostly tank penetrators, were used. Some 320 t of DU munitions were dropped in Kuwait and southern Iraq. Over one-third of combat veterans of the 1991 Gulf War showed an increased rate of tumors, immune disorders, and other symptoms, including chronic pain, fatigue, and memory loss.

Veterans involved in the Persian Gulf War and the Bosnia and Kosovo War have been found to have up to 14 times the usual level of chromosome abnormalities in their genes. A 2001 study (Kang et al. 2001) of 15,000 US Gulf War combat veterans and 15,000 control veterans found that the Gulf War veterans were 1.8 (male) to 2.8 (female) times more likely to have children with birth defects. Medical records 2 years later showed that the birth defect rate increased by more than 20 % (Wikipedia: http://en.wikipedia.org/wiki/Depleted_uranium#Health_considerations).

As for other symptoms such as chronic pains, fatigue, and mental disorders, the US Department of Veterans Affairs found no association between these symptoms and DU exposure, and concluded that "exposure to DU munitions is not likely a primary cause of Gulf War illness". Many other possible causes have been suggested, but they will not be pursued further here (Wikipedia: Gulf War Syndrome: http://en.wikipedia.org/wiki/Gulf_War_syndrome).

16.4.3 DU Effects in Yugoslavia (Bosnia, Herzegovina, Serbia)

The UN General Assembly published a report from Bosnia and Herzegovina about the Yugoslavia conflict on 2008.07.24. It reported that, in 1995–1999 and 2005, a total of 10,800 pieces of DU munitions were fired on Bosnia and Herzegovina, which amounted to approximately 2.9 t of uranium in the NATO bombings. No health effects were mentioned in this report, because, it says, the locations of contaminated areas were not accessible.

A documentary film ("Buried Warning", 2005) reports the health effects of these DU bombs. A tank manufacturing plant in Hadžići, Bosnia, was bombarded by 1,500 rounds of DU munitions in 1995. Three thousand workers at the plant moved to a small village, Bratunac, eastern Bosnia, where about one-third of them have died from various cancers since then. In a hospital in Sarajevo, the director told the reporters that cancer cases increased sharply by 4.5 times between 1996 and 2002.

Another report (October 27, 2009: http://rainbowwarrior2005.wordpress.com/2009/10/29/nato-bombings%E2%80%99-aftermath-takes-toll-on-serbia/) gives more details. Dr. Srbljak, head of the Merciful Angel NGO, says: "There is no other place in the modern world where so many people and so many young people – aged between 30 and 40 – die from cancer. Blood and lung cancer are most widespread." M. Milkovic, a veterinarian in southern Serbia, says: "Over the last 10 years, I have seen many two-headed calves, six- or eight-legged lambs and other anomalies among animals. Mutation is a normal thing, but there are so many cases – it's a symptom. Our nature is sick. And certainly – it has to do with depleted uranium usage".

Ibrulj et al. (2007) studied the peripheral lymphocyte genes in the people in Hadžići likely exposed to DU and those in a control area where no DU was used. They found a statistically significant increase in the chromosome aberration frequency in the exposed group in 2007. Micronuclei increased also in lymphocytes in the same group (Krunic et al. 2005). Micronuclei levels in peripheral blood lymphocytes have been demonstrated as a potential biomarker for pancreatic cancer risk (Chang et al. 2010) (Refer to Sect. 12.3.2.2 for micronuclei.)

16.4.4 Iraq/Fallujah/Basra

US troops almost destroyed Fallujah, west of Bagdad, in 2003–2004, and the population diminished by 30–50 % from the prewar period. Many birth defects have been reported in Fallujah since.

Robert Fisk describes many instances of congenitally malformed babies in his recent essay: "The Children of Fallujah – The Hospital of Horrors" (Fisk 2012). Just a few examples are: a baby with a hugely deformed mouth; a child with a defect of the spinal cord, material from the spine outside the body; a frog-like creature in which all the abdominal organs are trying to get outside the body; a tiny child with half a right arm, no left leg. Some example of deformed babies may be seen at http://www.youtube.com/watch?v=JC8uQp-FaS8.

Robert Fisk says: "this is too much. These photographs are too awful, the pain and emotion of them impossible to contemplate." Yet, more and more examples of birth defects were shown to him in a hospital in Fallujah. The doctors in the hospital told Fisk that they were cautious in identifying the causes for all these observations, but they say that it is alarming to see the increase in birth defects, miscarriages, and stillbirths in the area.

A most recent statistic for perinatal and neonatal mortality in Fallujah General Hospital has been reported by Abdulghani et al. (2012). The perinatal mortality during 2010 was 50.3/1,000 total live births, and the neonatal mortality rate (NMR) was 41.5/1,000 total live births. The NMR during 2007–2009 was 57.3/1,000 live births. One of the main reasons for the high mortality was low birth weight; 34.4 % were less than 1,500 g and 39.6 % had a birth weight in the range 1,500–2,500 g. The high NMR in Fallujah should be compared with the rate of 23/1,000 live births in Iraq as a whole, 4/1,000 in Qatar, 5/1,000 in UAE, 6/1,000 in Kuwait and Bahrain, 8/1,000 in Lebanon, and 9/1,000 in Libya in the same period. However, higher rates have been observed in Morocco, Djibouti, Sudan, and Afghanistan (ranging from 23 to 50/1,000).

The increase in birth defects has been reported for years since the bombing in 2003. Iraqi researchers and doctors have documented an increase of birth defects and cancer primarily in southern Iraq, where most of the fighting took place in the first Gulf War. It is spreading now, with the second Iraq war. DU has been singled out as the most likely cause (Johnson 2005).

Some authors suggest neurotoxic metals, lead (Pb) and mercury (Hg), as the cause of some of the ill effects seen in Iraq people (Al-Sabbak et al. 2012). Their study found that the population studied in Fallujah had been exposed to high levels of Pb and Hg. They found even higher levels of lead exposure in Basra. The authors note that toxic metals such as Hg and Pb are an integral part of war ammunitions and are extensively used in the making of bullets and bombs, and that the bombardment of Basra and Fallujah may have exacerbated public exposure to metals, possibly culminating in the current epidemic of birth defects. Others attributed the cause to white phosphorus used in another kind of ammunition. It is not quite reasonable to attribute the cause of birth defects to these other than radiation, because phosphorus and its derivatives or Pb or Hg is not known to cause malformation, though they may affect brain activities. High incidents of cancer have also been noted as well (Al-Faluji et al. 2012).

A recent report by Der Spiegel (December 18, 2012) depicts some observations of deformed babies and their deaths in Basra in southern Iraq. A cemetery manager said that there were several thousands of graves there, and another five to ten were added every day. There was a sevenfold increase in the number of birth defects in Basra between 1994 and 2003. About 100–200 t of DU ammunitions are believed to have been used in Basra, and the remains of tanks destroyed in the war with DU ammunitions littered the streets until 2008. Figure 16.7 shows photos of children

Fig. 16.7 Children playing on the destroyed (by DU munitions) tanks in Basra (http://www.ltu.se/cms_fs/1.85152!/file/2.3%20Almuqdadi%20and%20Alansari%201.pdf)

playing on the destroyed tanks in Basra (Al-Muqdadi and Al-Ansari), which were likely covered with powdery uranium oxide

16.5 An Overall Trend in Cancer Rates

Chapters 13 through 16 surveyed the adverse health effects of radiation associated with the obvious nuclear events, facilities, and accidents. The most studied ill effect was various cancers, including leukemia. The radioactive material spewed out from atomic bombs, their explosion tests, and nuclear power plants would be mostly in the form of fine to minute particles (aerosol), which can be carried by the air currents throughout the whole world, and could be washed down by rain. Figure 16.6 (observations in the UK of DU munitions used in the Middle East) and the observations with regard to the Chernobyl and Fukushima incidents have indicated that it is, indeed, the case.

The crucial data on radiation across this planet and its change over its history are not available. Figure 16.8 shows the world uranium production and demand. The total amount of uranium unearthed for human usage is shown by the blue line. It shows a sudden increase from 1950, and the amount of radioactive material produced would have been a multiple of the amount of uranium unearthed, as the fission reaction would have produced a lot of fission products with shorter (i.e., than uranium) half-lives.

How the overall radiation on this planet might have changed over the course of its history is indicated in Fig. 16.9. It simply depicts conceptually how the radiation on this planet might have changed over the earth's history, and, also, the likely change over the last century or so. The change caused by Chernobyl has been estimated to amount to about 2 % on average above the background level (Yablokov et al. 2009). The increase would have been much greater in the northern hemisphere alone though. Whatever is the real number, it is certainly true that the

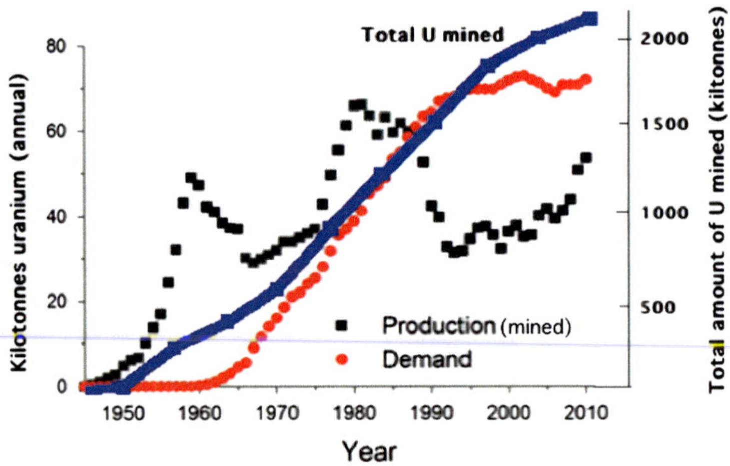

Fig. 16.8 World uranium production (mined) and demand (http://www.world-nuclear.org/info/inf23.html, modified from Wikipedia)

amount of radiation which all the creatures on earth are subject to has been suddenly increased by human activities in the last 50–100 years, as illustrated.

Sweden is located far from the atomic bomb dropping and test sites, though it is close to Chernobyl. A number of vital statistical data on cancers over the last century were analyzed by Hallberg and Johansson (2002) and the results are shown in the form of the temporal change of cancer/mortality rates. Figure 16.10 shows the change in the lung cancer death rate from 1910 to 2000. A sudden and rapid rise is seen from about 1950. The black lines are drawn by the original authors. Other cancers have also shown similar trends, as seen in Fig. 16.11. Very similar trends have been seen in many countries, including Japan, as shown in Fig. 13.6.

This increase in cancer is parallel to that in which radiation change surmised, as shown in Fig. 16.10, with some delays. This does not prove the causal relationship between cancer and the radiation, but it is an interesting coincidence. The other reasons offered for the increase in cancer in the last several decades are: (1) that the number of cancer-causing chemicals entering the environment has increased and (2) that cancers are typically a disease of old age and people have come to live longer, and, hence, the higher rate of cancer. It is unlikely that the spread and contamination of the environment by cancer-causing chemicals has suddenly changed around 1950.

As to reason (2), it might be pointed out that the increase in cancer in recent decades has been the highest among children and young people. A study by Steliarova-Foucher et al. (2004) found a general increase in cancers among children and adolescents across Europe. To begin with, it has to be understood that cancer has been rare before the age of 20 years. The investigators studied high-quality data from 63 European countries of 113,000 tumors in children and 18,243 tumors in

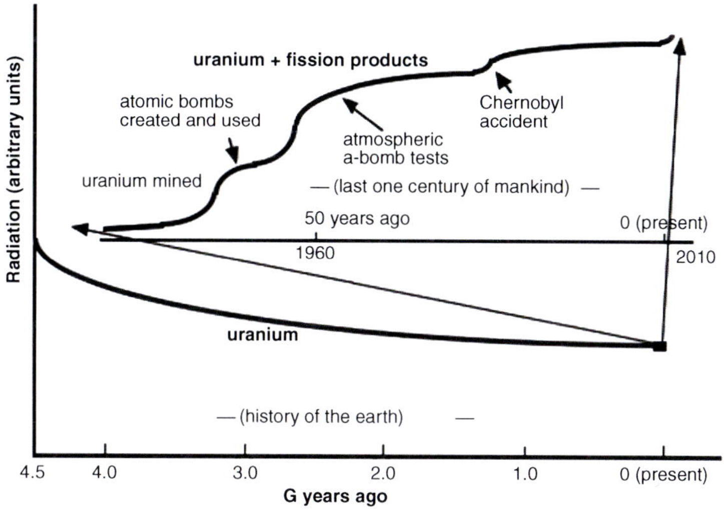

Fig. 16.9 Radiation level change on the earth (conceptual)

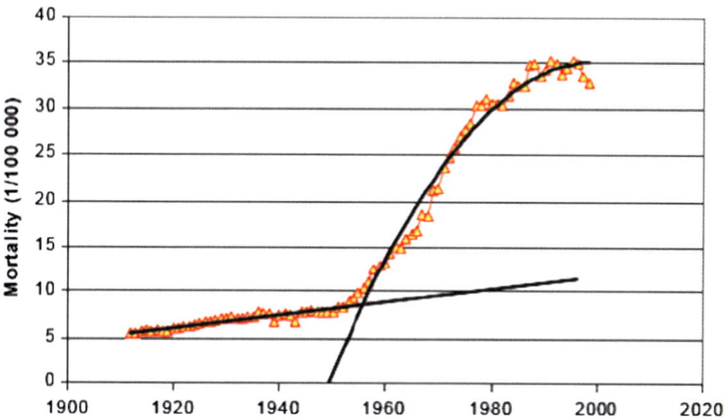

Fig. 16.10 Lung cancer death rates in Sweden (From Hallberg and Johansson 2002)

adolescents diagnosed in the period 1970–1999. Over the three decades, the overall incidence rate increased by 1.0 % per year ($p < 0.0001$) in children (aged 0–14 years) and by 1.5 % ($p < 0.0001$) in adolescents (aged 15–19 years). In children, increases were seen in most types of cancer, and prominent cancer rate increases were seen for carcinoma, lymphomas, and germ-cell tumors among adolescents.

Still, it cannot be proven unequivocally that radiation is the cause, but at least it might be said that radiation is a contributing factor. It might also be pointed out that the mortality rates of several cancers in Sweden at least seem to have flattened out beginning around 1970, implying an improvement in the treatment of cancer.

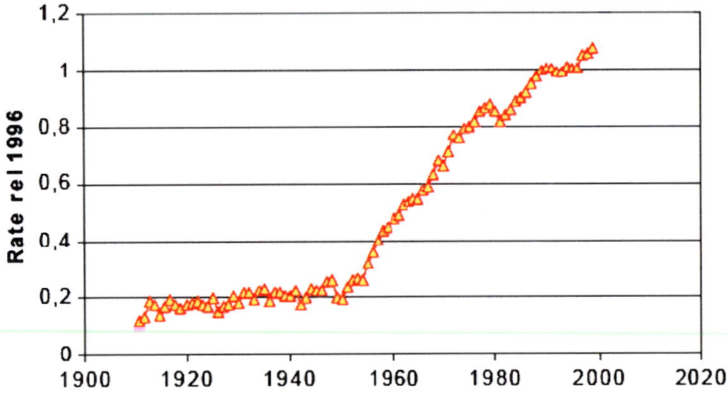

Fig. 16.11 The average cancer rate relative to 1996 for bladder, melanoma, prostate, lung, and breast cancers (From Hallberg and Johansson 2002)

However, at least in the case of breast cancer, the rate of incidence keeps increasing, despite the flattening out of the mortality rate.

References

Abdulghani ST, Sirhan YT, Lawas AK (2012) Perinatal and neonatal mortality in Fallujah General Hospital, Fallujah City, Anbar Province, west of Iraq. Health 4:597–600

Al-Faluji AAR, Aki SH, Al-Esawi AAJ (2012) Incidence of cancer in Fallujah above 10 years age with over view of common cancers in 2011. Health 4:591–596

Al-Sabbak M, Sadik Ali S, Savabi O, Svabi G, Dastgiri S, Savabieasfahani M (2012) Metal contamination and the epidemic of congenital birth defects in Iraqi cities. Bull Environ Contam Toxicol 89:937–944

Baker PJ, Hoel DG (2007) Meta-analysis of standardized incidence and mortality rates of childhood leukaemia in proximity to nuclear facilities. Eur J Cancer Care 16:355–363

Buried warnings: about the severe effects of DU bombs in Yugoslavia, a documentary film, Hiroshima Home TV (2005) http://ima-ikiteiruhushigi.cocolog-nifty.com/yugoslavia/2005/08/post_ccae.html

Chang P, Li Y, Li D (2010) Micronuclei levels in peripheral blood lymphocytes as a potential biomarker for pancreatic cancer risk. Carcinogenesis 32:210–215

Craft ES, Abu Qare AW, Flaherty MM, Garofolo MC, Rincavage HL, Abou-Donia MB (2004) Depleted and natural uranium: chemistry and toxicological effects. J Toxicol Environ Health Part B Crit Rev 7:297–317

Durakovic A, Gerdes A, Zimmerman I (2004) Concentration and ratio of uranium isotopes in the fine-fraction of surface soil from Baghdad and Basra collected after Operation Iraq Freedom. http://umrc.net/wp-content/uploads/2012/06/Concentration-and-Ratio-of-Uranium-Isotopes-in-the-Fine-Fraction-of-Surface-Soil-from-Baghdad-and-Basra-Poster-Asaf-Durakovic-HPS-2004.pdf

Fisk R (2012) The Children of Fallujah – the hospital of horrors. The Independent, 27 April 2012

Gould JM, Goldman BA (with Millpointer K) (1990) Deadly deceit, low-level radiation and high-level coverup. New York: Four Walls Eight Windows

Gould J (1996) The enemy within, the high cost of living near nuclear reactors. Four Walls Eight Windows, New York

Hallberg Ö, Johansson O (2002) Cancer trends during the 20th century. J Aus Coll Nutr Environ Med 21:3–8

Huffington Post (2011) Radioactive tritium has leaked from three-quarters of U.S. nuclear plants: AP investigation. http://www.huffingtonpost.com/2011/06/21/radioactive-tritium-leaks-us-nuclear-plants_n_881090.html

Ibrulj S, Haverić S, Haverić A (2007) Chromosome aberrations as bioindicators of environmental genotoxicity. Bosnian J Basic Med Sci, 7:311–316, cited in ECRR 2010

Joel K-H, Greiser E, Hoffmann W (2007) http://www.bfs.de/de/kernetechnik/kinderkrebs/Qualitaetspruefung.html

Johnson L (2005) Doctors report "unprecedented" rise in deformities, cancers in Iraq. http://rainbowwarrior2005.wordpress.com/tag/deformed-babies/

Kaatsch P, Spix C, Schulze-Rath K (2008) Leukaemia in young children living in the vicinity of German nuclear power plants. Int J Cancer 122:721–726

Kang H, Magee C, Mahan C, Lee K, Murphy F, Jackson L, Matanoski R (2001) Pregnancy outcomes among US Gulf War veterans: a population-based survey of 30,000 veterans. Ann Epidemiol 11:504–511

Krunić A, Haverić S, Ibrulj S (2005) Micronuclei frequencies in peripheral blood lymphocytes of individuals exposed to depleted uranium. Arh Hig Rada Toksikol, 56: 227–232, cited in ECRR 2010

Mangano JJ (2009) Geographic variation in US thyroid cancer incidence and a cluster near nuclear reactors in New Jersey, New York, and Pennsylvania. Int J Health Serv Plann Admin Eval 39:643

Nussbaum RH (2009) Childhood leukemia and cancers near German nuclear reactors: significance, context, and ramifications of recent studies. Int J Occup Environ Health 15:318–323

Sauer JR (2011) Health concerns and data around the Illinois nuclear power plants. http://dels.nas.edu/resources/static-assets/nrsb/miscellaneous/Sauer_morning_present.pdf

Sermage-Faure D, Laurier D, Goujon-Bellec D, Chartier M, Guyot-Goubin A, Rundant J. Hémon D, Clavel J (2012) Childhood leukemia around French nuclear power plants – the Geocap study 2002–2007. Int J Cancer, 131: 769–780, first publ. Feb 2012 (e-pub)

Steliarova-Foucher E, Stiller C, Kaatsch P, Berrino F, Coebergh J-W, Lacour R, Perkin M (2004) Geographical patterns and time trends of cancer incidence and survival among children and adolescents in Europe since the 1970s (the ACCIS project): an epidemiological study. Lancet 364:2097–2105

Sternglass E (1972/1981) Secret fallout: low level radiation from Hiroshima to Three Mile Island. McGraw-Hill, New York

Yablokov AV, Nestrenko VB, Nestrenko AV (2009) Chernobyl: consequences of the catastrophe for people and the environment. Ann NY Acad Sci 1181:E1–E39

Chapter 17
(High-Energy) Radiation Is Incompatible with Life

Chapters 6 though 16 considered the interaction of high-energy radiation with chemical entities, atoms, molecules, and compounds, particularly those found in biological systems, and the resulting health effects and available defense mechanisms, including repairing DNA damage. All of the scientific considerations and data accumulated so far seem to indicate the basic tenet that "High-Energy Radiation is Incompatible with Life".

However, organisms have lived over the last four billion years on this planet, where a background level of radiation has persisted. That means that organisms have survived despite the background radiation, and it was because they have developed a few mechanisms that could make them somewhat tolerant to the damaging effects of radiation. How effective the repair mechanisms are in combating the radiation effects has not yet been delineated, but the organisms seem to be capable of just barely coping. It has been demonstrated in this treatment (Sect. 12.5) that it is, indeed, the case, so a sudden rise in the radiation due to human activities seems to have caused widespread adverse health effects. If the coping mechanism was relatively sturdy and may tolerate several times the original background level of radiation, the tragedy, i.e., the adverse effects of radiation, of Chernobyl and others would not have happened. A recent study has, indeed, demonstrated that about 15 % of leukemia among children in the UK in recent decades is due to the background radiation (Kendall et al. 2012/2013).

17.1 Incompatibility of Radiation with Life

Radiations such as X-, α-, β-, and γ-rays have large energies, ranging from several thousands to millions of times the energies associated with chemicals and chemical reactions. The exception is ultraviolet light, which is of the same or, at most, several times as large as the chemical energy range, and, typically, it excites the electronic states in them, and would not ionize, as the other radiations of a higher energy do.

E. Ochiai, *Hiroshima to Fukushima*, Science Policy Reports,
DOI 10.1007/978-3-642-38727-2_17, © Springer-Verlag Berlin Heidelberg 2014

The impact on atoms and molecules of these high-energy particles (γ- and X-rays are regarded as particles (photons)) is to kick out electrons, break bonds, deform/decompose molecules, and form free radicals. The energy of a single particle is at least several thousands of times of that required to effect these chemical changes, and a single impact would usually reduce only a small portion of its energy. Hence, a single high-energy radioactive particle would impact on several hundreds to thousands of molecules located nearby. An α-particle, because of its high electric charge and mass (mass number 4), would give high impacts at high frequency, but be retarded of its speed relatively quickly, and, besides, it loses its electric charge as it captures electrons as it slows down. So, its distance of movement among molecules and bodily organs is the shortest among the radioactive particles, but the density of affected entities is high. A β-particle, being a negatively charged electron, goes farther, and the electromagnetic wave, γ-particle, travels farther; typically, it goes through an entire human body, as detailed in earlier chapters.

The destruction of molecules such as that caused by high-energy radiation would not happen in ordinary chemical reactions. Ionization and bond breaking can be brought about by chemical means within chemically feasible energy, and, hence, can be controlled within the means of chemically available energy. But destruction by high-energy radiation cannot be regulated by the chemical compounds and chemical energies available. This is the basis for the contention that high-energy radiation is incompatible with the chemical world, life in particular, as life is entirely of chemical phenomena and very vulnerable to any disturbance.

In living organisms, various inherent mechanisms do exist to make corrections, mend, and repair some damage on biomolecules, cells, and some tissues. These functions have evolved, not necessarily for the purpose of repairing radiation damage, but for repairing the damage caused by biological/chemical actions on biological/chemical entities. Therefore, most of the damage and ill conditions caused by chemical and biological activities (including disease-causing bacteria and viruses) may be repaired and corrected by those means developed throughout biological evolution, particularly through the invention of immunity. This is a theoretical argument, but the reality is that humankind has devised remedies for some but not all disturbances. There are elaborate systems to repair wrong sequences in DNA, damaged DNA, mostly caused by biological and chemical agents. However, some of these mechanisms may be applied to repair some damage caused by radiation. The dimerization of adjacent thymines can be caused by ultraviolet light, though radiation of higher energy would not do this. This occurs through an electronic excitation of the $C = C$ bonds on the adjacent thymine residues. A specific repair mechanism for this dimer has evolved, as outlined earlier (Sect. 12.2.2). Some damage (DSBs on DNA) seems to be unrepairable by the endogenous mechanisms as mentioned in Sect. 12.3.4.5.

The so-called reactive oxygen species (ROS) include the excited singlet state of dioxygen, peroxide (ROOH), hydrogen peroxide (HOOH), superoxide radical ($O_2\bullet^-$), and hydroxyl radical (HO•). A few biological entities/enzymes are present to reduce the reactivities of ROSs (Sect. 12.1.2, pp. 282–284 in Ochiai 2008).

Superoxide dismutase (an enzyme) converts superoxide to dioxygen and a less severe hydrogen peroxide, which can be decomposed by catalase. A number of peroxidases decompose hydroperoxides, as discussed earlier. However, no specific enzyme is known to deal with hydroxyl radicals, though some biocompounds such as ascorbic acid (vitamin C), flavin, polyphenol, and some thiols can kill the free radical character of the hydroxyl radical and other free radicals. These biocompounds are not developed specifically for combating the effects of radiation, though they would be able to reduce, to a certain degree, the effects of free radicals created by radiation.

Radiation likely destroys almost any chemical compound encountered in the biological systems. For such destructions, there are not enough endogenous repair mechanisms. For example, diarrhea can be caused by many (bio)chemical factors and can be treated by medicines (chemical compounds) once the cause is diagnosed. However, it turned out that no remedy (medicine) has been found to be effective in healing diarrhea caused by radiation. The therapeutic use of X-rays in the abdomen is known to often cause diarrhea. Radiation damages the membrane structures, mitochondria, and cytoskeleton, as well as nucleic acids in enterocytes, and results in denudation of the basal membrane, crypts, and epithelial villi in 3–4 days; this is the cause of diarrhea (e.g., Blanarova et al. 2009) and it is difficult to repair promptly by chemical and biological means. The enterocytes are supposed to renew every half a day, but radiation stops them from doing so. Postradiation inflammatory changes are nonspecific, however, and difficult to distinguish from inflammation resulting from other diseases. Attempts to treat the complications caused by radiation-induced enteritis and colitis have been unsuccessful; antibiotics, sucralfate, anti-inflammatory drugs, octreotide, proteolytic enzymes, etc. were tried.

This is just an example showing how different the radiation damages are compared to the ordinary damages caused by chemicals and biological agents. Chemical and biological means to defend against damage are limited, and may not extend to some of the cases caused by radiation, as living organisms have not encountered them before the increase of the radiation levels on the earth. The sudden increase in the amount of radiation on this planet has taken place only at most in the last 100 years. Living organisms are not yet prepared enough to cope with such a devastating and sudden assault of high-energy radiation.

There seems to be rather elaborate mechanisms to repair a number of defects and damages in DNA, as DNA is the basic crucial entity in life. Some of these mechanisms are utilized to repair damage caused by radiation. Henceforth, the effects of low-level radiation would least likely manifest in cancers. This suggests that other adverse health effects such as diarrhea, cardiovascular disease, brain damage, deformed babies, chronic fatigue syndrome, weakening of the immune system, etc. would be more prevalent than cancers. Most data from the Chernobyl incident indicate that this is, indeed, the case. The current situation following the Fukushima incident suggests that it is, indeed, the case, but no definitive data are yet available. Besides, all human diseases may be caused by many other factors, and, henceforth, it is quite difficult to establish the causality between a disease and radiation.

Unfortunately, the "authorities" (UNSCEAR, IAEA, ICRP, WHO) in the nuclear industry have recognized only the cancers, specifically, thyroid cancer and leukemia, as the result of radiation. However, the authoritative RERF for the original data on the radiation effect (which has been based on the survivors of the atomic bombs) has now included all types of cancer in their LSS-14 report, as seen in Sect. 13.2.3, acknowledging implicitly that all types of cancers could be caused by radiation (from the atomic bombs). The LSS-14 report also deals with the noncancerous diseases versus dose relationships, implying that they admit that those diseases may be to a certain degree caused by radiation.

17.2 Radiation in the Evolution of Life

The incompatibility of high-energy radiation with life can also be illustrated by the way that life on this planet has evolved. When the earth was formed, there still remained a significant quantity of radioactive isotopes, and no significant amount of free oxygen (O_2) was present in the atmosphere. Many of those radioactive isotopes disappeared in the earlier stages of the history of the earth, as their half-lives were not very long, and, hence, only the radioactive isotopes of long half-lives remained, and some of them are present even today. These isotopes were mentioned previously in this treatise. That is, it should be recognized that the radiation on this planet decreased significantly as time went by, but it still remains today as a part of the background radiation, as mentioned earlier.

Free oxygen (O_2) in the atmosphere, if present, will form ozone (O_3). O_2 as well as N_2, the major components of today's atmosphere, do not cut off the ultraviolet component of sunlight. So, early in the history of this planet when no free oxygen was present in the atmosphere, the earth's surface was showered with ultraviolet light and cosmic rays, and was not conducive to life, as the ultraviolet light (as well as the cosmic rays) was toxic to life, as recognized by most people today. The electrically charged particles such as α and β (of cosmic rays) from the sun are deflected by the geomagnetism, which is believed to have been established by 2.7–2.8 billions ago, and not much would have reached the surface of the earth.

Life started its long journey on the planet in the depths of the oceans, where life-supporting substances were spewed out from the cracks of the ocean floor. They were subject to radiation of low level in the seawater, but not cosmic rays or ultraviolet light. Cyanobacteria are believed to have emerged 2.5–3 billion years ago and came to shallow waters to harvest sunlight after most of the cosmic rays had been shielded off, developing photosynthesis, thereby, decomposing water through sunlight. Photosynthesis produces free oxygen (O_2).

Enough free oxygen accumulated in the atmosphere by around 2.2 billion ago and it had formed an ozone layer in the upper portion of the atmosphere by about 0.4–0.5 billion years ago. The toxic ultraviolet light from the sun was cut off and not much of it reached the surface of the earth. This condition allowed living organisms to emerge from the oceans and onto land. This fact suggests that, even though photonic energy is relatively low, a high dose of ultraviolet radiation is lethal to organisms. It would further suggest that particles of higher energy, α, β, and γ, would be lethal at high levels, but would also cause adverse effects even at low levels. It might be pointed out that a high or low dose of radiation has nothing to do with the strength of radiation, but, rather, represents the number of radiation particles. The strength of radiation is dependent on the kinetic energy of the radiation particle. In other words, low-level radiation would produce the same kind of effects on biomolecules as high-level radiation, only that the number of affected sites would be smaller.

By the way, free oxygen is toxic to living organisms and many organisms did not survive when the amount of atmospheric oxygen became significant. Many organisms devised means to make use of the free oxygen (by developing oxidative metabolism) and simultaneously reducing the level of toxic free oxygen in their bodies, and, also, they devised many enzymes and others to detoxify oxygen and its active derivatives.

Until a century or so ago when the human race started to produce a number of artificially created chemicals, the environment had only naturally occurring chemicals, with which many living organisms had devised coping mechanisms. The radiation in the background was and still is causing disturbances in the chemical world of life. Organisms had to cope with the radiation as well; otherwise, they would not have survived. Luckily, the background radiation level was not very high, and a few mechanisms to repair DNA and other mechanisms to detoxify some substances produced by radiation seem to have been able to cope with it, but only barely. The radiation effects would have caused mutations in the genes, and most of them were deleterious, and some organisms must have died through its adverse effects. However, occasionally, some changes in the DNA might have led to new species. This is only a very rare occurrence, but it might have been one of the causes for the creation of new species, i.e., evolution, though it must be said that the mechanism of species creation has not yet been fully understood.

17.3 Difficulties in Combating Radiation Effects in Real Life

Another reason for incompatibility is not a scientific but a practical issue. Humans cannot detect nor measure radiation intensity by their senses alone; no smell, no vision, no touch sense, no sound. One can only know the presence of radiation by a detector and measuring device. The workers who deal with radioactive material or radiation devices usually carry an instrument to measure the exposed dose, to monitor their own exposure, wear protective clothing, and control their work conditions accordingly. Unfortunately, this practice is not enforced in many situations.

Ordinary people do not take such precautions. The officials responsible would publish the distribution maps of radioactive material (Bq/m^2 for Cs-137 for example) or radiation dose (Sv/hr), so that people could avoid living in contaminated areas. However, some people cannot afford to or would not move away for one reason or another. Those people who stay in contaminated areas cannot know how much they would be exposed to radiation or how contaminated are the food or drink which they consume. Are the people who think that they live outside of contaminated areas safe? How about what they eat? Are foods that are contaminated at lower levels than the regulation safe enough? What are the bases for the safety levels anyway?

Decontamination means not only removing radioactive material but also to safely dispose of the material thus removed. First, there are a number of situations in which the removal of contaminated material is difficult in practice. An example is forest where radioactive material will be inside the trees themselves or on the surface, soil, fallen leaves, etc. Even if the radioactive material has been completely removed from one area, there are always chances that radioactive material can come flying or flowing back to the area. In other words, the decontamination of an area can be effective only temporarily, but cannot be ideal for a long time. People's lives cannot be adjusted accordingly. This comment is not intended to deny the usefulness of well-managed decontamination.

Decontamination of the affected bodies is the crucial and difficult issue, but this is the only effective means to protect the human body from radiation. Currently, apple pectin seems to be the only agent used for this purpose (Sect. 13.2 in Yablokov et al. 2009).

Pectin is typically a mixture of polysaccharides and their derivatives, rich in galacturonic acid, and is found in primary plant cell walls and is abundant in the nonwoody parts of terrestrial plants. The carboxylate groups and some hydroxyl groups abound in pectin bind and wrap cations, including Cs(+1) and Sr(+2). They are called "enterosorbents", meaning absorbent in the intestine. It seems to be effective to sorb (wrap up) Cs(+1) (and others) and increase its fecal excretion.

A study on children in Svetlogorsk city in Gomel province, Belarus, showed that 615 children who took a pectin formula (5 g twice a day for 3 weeks) along with clean food had their Cs concentration reduced by 63 %, as compared to 14 % in children given a placebo. Another study found that the mean effective half-life of Cs137 was 27 days with pectin, but 69 days without pectin (Nesterenko et al. 2004; Sect. 13.2 in Yablokov et al. 2009). These studies indicate that pectin is effective in excreting Cs-137. Since pectin is not readily absorbed from the gastrointestinal tract, it binds Cs only while it is in the gastrointestinal tract and is then excreted along with feces. Once Cs or some other radioactive entity is absorbed through the gastrointestinal tract and into the circulatory system, pectin may not be effective.

Prussian blue is another agent to help remove Cs (http://www.bt.cdc.gov/radiation/prussianblue.asp). This also acts in the intestine, where it binds Cs and the complex is excreted with feces. The soluble form of Prussian blue is potassium salt, but potassium can be exchanged with Cs, forming an insoluble compound, which can be excreted as such.

It is imperative that the scientific community tries to find effective means to remove the radioactive substances from a body, though it would be difficult. The quantity of radioactive material in a typical internal exposure is so small that it would not be easy to find a very specific and effective agent to bind only that element. Then, the combined entity has to be effectively and rapidly excreted from the body. As illustrated above, the removal of a radioactive substance present in the intestinal tract is relatively easy, but the removal of those present in circulating blood, tissues/organs/cells are not so easy. The target radioactive isotopes include Cs-137, Sr-90, and many others (see Table 2.1). Nonradioactive iodine compounds such as KI is prescribed for protecting the thyroid gland from radioactive iodine (I-131, 129). Iodine is a component of the thyroid hormone thyroxine, and, hence, it is readily absorbed into the thyroid. The nonradioactive iodine (KI) is meant to dilute the iodine in the body, so that the absorption of radioactive iodine may be reduced. This is not decontamination.

References

Blanarova C, Galovicova A, Petrasova D (2009) Use of probiotics for prevention of radiation-induced diarrhea. Bratisl Lek Listy 110:98–104

Kendall G, Little MP, Wakeford R, Bunch KJ, Miles JCH, Vincent TJ, Meara JR, Murphy MFG (2012/2013) A record-based case-control study of natural background radiation and the incidence of childhood leukaemia and other cancers in Great Britain during 1980–2006. Leukemia 27(2013):3–9 (Published online 6 June 2012). http://www.nature.com/leu/journal/v27/n1/abs/leu2012151a.html

Nesterenko VB, Nesterenko AV, Babenko VI, Yerkovich TV, Babenko IV (2004) Reducing the 137Cs-load in the organism of "Chernobyl" children with apple-pectin. Swiss Med Wkly 134: 24–27

Ochiai E (2008) Bioinorganic chemistry: a survey. Elsevier, Amsterdam

Yablokov AV, Nestrenko VB, Nestrenko AV (2009) Chernobyl: consequences of the catastrophe for people and the environment. Ann NY Acad Sci 1181:E1–E39

Part VI
The Science of Biohazards of Radiation Has Been Influenced by Non-scientific Interests

Radiation biology was initially concerned with X-rays, particularly their diagnostic use. The radiation comes from outside of the body in this case. Next came the radiation from radium and other radioactive material, and the concern was still the external exposure. When the atomic bomb was developed and eventually used on human beings, the concerns were the physical (heat and wind) effects and the severe external radiation. Some scientists, however, seem to have been aware of the possibility of internal exposure effects from the fallout, but the military and government ignored or tried to suppress this idea. The commercial application of nuclear fission reaction, in the form of nuclear reactors, was initially thought to be safe and comprise radiation-free operations, despite the technical difficulties and complexity involved. Numerous nuclear weapon tests and the release of radioactive material from nuclear facilities, accidental or otherwise, have suddenly increased the amount of radiation in the environment. This has caused serious health problems, particularly through internal exposure to radiation. However, the authorities associated with the nuclear industry have been, and still are, very reluctant to recognize internal exposure and the reality of the health effects.

Chapter 18
Falsification and Suppression of the Truth

18.1 Development of the Idea of "Radiation Biology"

The radiation phenomena were discovered towards the end of the nineteenth century, consecutively in the form of X-rays by W. C. Roentgen (1885), uranium radiation by H. Becquerel (1886), and radium radiation by M. S. and P. E. Curie (1888).

In the 1880s, patients who had X-ray photographs taken of their bodies experienced some ill effects, such as hair loss, skin redness, desquamation, etc. Some scientists and medical doctors suggested caution against the radiation effects of X-rays. However, the prevailing sense was that the use of X-rays was safe without ill effects, and had ignored these cautions, until C. Dally, an assistant of Thomas Edison, died from high dose exposure to X-rays. He was involved in developing X-ray apparatus and suffered burns, serial amputations, and extensive lymph node anomalies, and, eventually, carcinoma, and died in 1904. This alerted the medical community to the issue of health effects of X-rays, and made Edison give up on the X-ray apparatus (pp. 5–6 in Forshier 2009).

Along with the discoveries of radiation from uranium and radium, this kind of incident sparked an interest in the biological effects of exposure to radiation. It was demonstrated by experiments based on X-ray irradiation on rodent testicles and others that stem or immature cells are more sensitive to radiation than mature cells, and cells that are metabolically active or growing fast are affected by radiation more severely than otherwise. By the 1920s, researchers formulated a basic mechanism of radiation; that is, radiation causes the ionization of molecules, directly or indirectly (through the formation of free radicals), as discussed in Part IV.

Subsequent developments were mostly concerned with the diagnostic use of radiation, radiation therapy and how radiation (X-rays) can be effectively applied, and how to protect the operators of X-ray instruments from the radiation effects, as well as the effects on patients. Concerns with the latter issue prompted the formation of regulatory agencies such as the International Commission on Radiological Protection (ICRP) in 1928 and the National Council on Radiation Protection and Measurements

E. Ochiai, *Hiroshima to Fukushima*, Science Policy Reports,
DOI 10.1007/978-3-642-38727-2_18, © Springer-Verlag Berlin Heidelberg 2014

(NCRP) in 1964, which later changed to the Nuclear Regulatory Commission (NRC). Throughout this development period (up to a few decades after the first atomic bombs), the concern was irradiation from the outside of the body, i.e., from a solid X-ray instrument and, later, the atomic bomb; this is essentially "external exposure". As a result of this tradition, the concern expressed by the authorities even today seems to be focused on "external exposure" and they tend to ignore "internal exposure" or belittle the internal exposure effects.

Today's major radiation problem, particularly caused by the accidents at NPPs, is "low-level internal exposure", but it is not above the horizon in the eyes of the authorities. Or, rather, it seems to have been suppressed by the authorities. We will sketch how this tendency has developed.

18.2 Manhattan Project up to Hiroshima/Nagasaki and Subsequent Developments

The Manhattan Project was strictly confidential. As it developed, a number of sectors involved had become infatuated with it in many senses. Scientists and engineers were very eager to pursue the very challenging works, the first in human history. The military officers who led the project were eager to get hold of the ultimate weapon and test and use it. The corporations involved were quite happy with the relationships with the military, the entitlements associated, and the resultant financial gains (from the government).

Scientists and engineers lived in an isolated and secured area, literally an atomic village. Academic departments such as nuclear physics and nuclear engineering flourished, attracting many brilliant young people, once the nuclear industry had established itself. Some aspects of nuclear development, including the rise of physicists, are detailed by S. Cooke (2009).

How the concern or neglect about radiation had developed among the personnel involved in the project has been well researched by H. Takahashi (2008, 2013). The following outline is based on her work.

There was some concern with radiation effects in the course of the development of nuclear weapons. Hence, a "radioactive poisons subcommittee" was created as part of the Manhattan Project in May 1943. The members included James B. Conant and Arthur H. Compton, and Conant, in his report of August 1943, stated that it would be fatal if only a millionth of a gram of a radioactive material (meaning plutonium) is deposited in a lung (Bush–Conant file, 1940–1945). That is, at least some of the scientists involved were aware of the internal radiation effects.

At a news conference on September 12, 1945, a month after the Hiroshima explosion, Thomas Farrell, the deputy director of the project, categorically denied that the radiation from the explosion had any ill health effects, though many people had died from the explosion itself (heat, blast, and immediate radiation by γ-rays and neutrons). He and his advisor, Stafford Warren, believed that the fireball formed during the explosion would have raised the radioactive material up into the stratosphere, and it would be blown away, diluted and spread far and wide, and, hence, no significant amounts would have come back down onto the explosion site and nearby. Warren measured the radioactivity in Hiroshima after the very strong typhoon (Makurazaki Typhoon on August 19, 1945) and found it to be very low. Apparently, most of the radioactive material had been washed away by the typhoon. But this result and his view on radiation effects have become the official governmental view thereafter.

Many Japanese scientists and medical doctors were gathered in Hiroshima and Nagasaki immediately after the incidents, and they collected a large amount of data about the casualties and diseases, including those among the people who were not in town at the time of explosion but came into town days or weeks later for one reason or another. Unfortunately, their reports were hidden in the US military archives and classified.

The US military began to investigate, with the cooperation of Japanese scientists, the casualties caused by the atomic bombs after they started to occupy Japan. In 1947, the Atomic Bomb Casualty Commission (ABCC) took over such an investigation, under the auspices of the US National Academy of Sciences. In 1950, the Atomic Energy Commission (AEC), Department of Defense, and Los Alamos Laboratory coauthored "The Effects of Atomic Weapons" (AEC et al. 1950). They asserted in the book that, because the explosions took place high above the ground in both Hiroshima and Nagasaki, the radioactive material was blown away, and, hence, no ill effects due to the fallout were reported.

However, as outlined in Sects. 11.3.4, 11.4.2, and 11.4.4, the large number of cases which suggested the ill effects of the fallout made the ABCC aware of the necessity of investigation of such cases, and formed the ME-81 branch to do so. It operated in 1952–1953, but was discontinued in 1953. The hydrogen bomb experiment at Bikini Atoll took into consideration the radiation effects on animals, and a project called the "Sunshine Project" was founded to explore internal exposure effects, collecting bones and urine and other samples from around the world. Yet, no official announcement included the findings of such an investigation on "internal effects".

The main work of the ABCC was still focused on the effects of immediate and external radiation due to the explosion of bombs. They started to follow the fate (death and cause) of the survivors in 1958, and the work was transferred to the Radiation Effects Research Foundation (RERF) in Hiroshima in 1975, and has continued as indicated by the LSS-14 report discussed in Sects. 13.2.1 and 13.2.2. In all the continuing investigations, the effects of the fallout, both internal and external exposure, have been ignored. This policy (or attitude) underestimates the low-level dose, particularly in the case of internal exposure, and would not give adequate measures for the risk to health by low-level radiation. Yet, this set of data is now considered to be the authoritative basic data to relate the exposure dose to health effects. Even today, the RERF would not take "internal exposure" into consideration in evaluating the dose effects on various cancers, even though many researchers have asserted the necessity to do so, as revealed, for example, by Sawada (2011, Sect. 13.1). Yet, the most basic data on radiation effects in the LSS-14 report now admits the linear with no threshold (LNT) idea for cancers, and, also, noncancerous diseases being caused by radiation. This fact is not taken account of in documents and policies by the most influential organizations: International Atomic Energy Agency (IAEA), ICRP, and the World Health Organization (WHO).

The United Nations Scientific Committee on the Effects of Atomic Radiation (UNSCEAR) was established in 1955, under the influence of the United States AEC. A report of this committee in 1958 included a number of findings about radioactive strontium present in various foods and the level of strontium in the bones of children in the samples collected from all over the world. Yet, the report only stated these findings and did not provide any recommendations.

An ultimate strategy to suppress the truth about the effects of radiation was to bring the UN organization WHO under the influence of the nuclear industry complex. The ICRP organized an unofficial gathering of all the agencies concerned: IAEA, UNSCEAR, WHO, UNESCO, ILO, FAO, and a few others in Switzerland in 1958. In 1959, the WHO made an agreement with the IAEA, in which the WHO would be required to obtain permission from the IAEA in order to publish anything associated with radiation issues due to the nuclear and its related industries.

18.3 Atomic Bomb Tests

An accidental finding of the reality of fallout from the atomic bomb tests took place in 1953 in Troy-Albany and its surrounding area in New York State, which is far away from the test site in Nevada (3,700 km), as mentioned in Sect. 16.1 (Sternglass 1972, p. 81). The radiation readings in the area turned out to be as high as thousands of times above the background level. The scientists from the AEC conducted an extensive study with better equipment. The detailed report of their findings was classified as "secret" and was never made public.

R. Lapp published an article about the fallout from the Nevada test in Science magazine in 1962 (Lapp 1962, 1963). He suggested that the Troy-Albany incident would provide a good opportunity to study the effects of fallout radiation on the health of people in the area, particularly thyroid cancer and leukemia in children, because there were enough data on radiation readings and enough members of the cohort, i.e., the affected people. However, the AEC and other agents ignored his suggestion or did not make the findings public. The fact that low-level ionizing radiation (X-rays) causes leukemia in children born from mothers who had been exposed to diagnostic X-rays in the abdomen (Stewart 1958) had been well publicized and known by that time.

A set of vital statistics of children under 15 years old in the Troy-Albany area for the period 1952–1962 was published in a letter to Science magazine in 1964 by an official of the New York State Department of Health. The letter concluded: "the cancer report files of this department reveal no increase in the incidence of cancer or leukemia over the past ten years in children of the Troy-Albany-Schenectady area (TAS) – who were 15 years old or younger in 1963 as compared with children of this age elsewhere in upstate New York" (cited from Sternglass 1972/1981). However, Sternglass points out that leukemia cases among children under 10 years of age in the TAS area numbered 9 during 1952–1955 (before the effects of fallout from the Nevada tests showed up) but increased to 32 during the peak period 1959–1962 (6–9 years after the fallout). The probability that this increase is purely accidental has been estimated to be one in a billion. The official's deception was that the comparison was made between the TAS area and the other area of upstate New York. The latter area seems to have also been contaminated by the fallout, and, hence, no significant difference was to be found. This is basically the same strategy for deception used in calculating the extra risk rate (ERR) among Hiroshima/Nagasaki survivors with reference to the control group who were significantly exposed and, yet, assumed to have had no significant dose.

18.4 Three Mile Island Incident

A case history of the suppression of significant data on the Three Mile Island incident has been described in detail by Sternglass (1972/1981). He examined the official vital statistical data provided to him by the officials and saw a significant increase in infant deaths/stillbirths soon after the accident. He knew from past experience that it was likely due to radioactive isotopes released from the accident, which got into, by inhalation, pregnant women and affected the fetus in the womb. He was invited to present his findings at a news conference in Harrisburg in November 1979. Before him, Dr. C. Kepford spoke about high radiation readings as a result of the accident. He had been fired from Pennsylvania State University because of his criticism about the Three Mile Island facility's release of radioactive material even before the accident. Sternglass spoke about the sharp rises in infant mortality in the areas surrounding the facility, but also as far away as Pittsburgh. What happened after the news conference is quoted below from his publication, "Secret Fallout" (1972/1981).

> Television cameras representing the major networks had been present; some of the network reporters interviewed me separately immediately following the news conference. But neither that evening nor the next day was there any mention of these disturbing findings either on the local news in Pittsburgh nor any of the national television programs. There were a few very brief radio news items, but not a word of the news conference appeared in any Pittsburgh or Philadelphia papers. It was as if an iron curtain had descended around the Harrisburg area, sealing off the people of the rest of the United States and the world from the news that would have warned them of a totally unexpected severe effect of low-level fallout. But neither the nuclear industry, the military, nor the state and federal governments committed to nuclear power wanted them to know. . . .

Then, efforts by the authorities and the mass media, including the New York Times, had followed to discredit the findings or destroy the credibility of the speaker. However, his findings and conclusions were verified by several other parties, and the authorities' tactics to mislead the people by picking and choosing data intentionally had been revealed by some. Yet, the official stance (no significant ill health effects) has persisted.

18.5 Chernobyl Incident

The government at the time of the accident was that of the Soviet Union and was in dire conditions, both politically and financially. It could not afford to adequately deal with the accident and its aftermath. It wanted to minimize the damage to the government by attempting to make the effects of the accident look not too serious. It suppressed the truth: the seriousness of the health effects on liquidators and the contamination by radioactive fallout and its effects on people's health.

Once the Soviet government fell and was dissolved, the responsibility for the accident and its aftermath came under the jurisdiction of individual states: Ukraine, Belarus, and Russia. And the international organizations WHO, UNSCEAR, IAEA, and ICRP moved into the position of mediating the various issues associated with the accident.

How these organizations, including the official departments of Belarus, Ukraine, and Russia, have influenced, mostly suppressed, the investigations and falsified reportings by various individuals and medical facilities in these countries are somewhat summarized in Chap. 8 of Pflugbeil et al. (2011). The reality of the deception is best illustrated by a documentary film on the 2001 Kiev Conference on the Chernobyl accident, titled "Where is the Truth?" (Andreoli et al. 2004).

It was revealed that the WHO, the organization protecting the world's citizens from ill health effects, had not visited the Chernobyl area for the first 5 years after the disaster, and, instead, it had assigned the IAEA to investigate it. The IAEA is the organization to oversee the nuclear industry. A. Yablokov, V. Nestrenko, Y. Bandazhevsky, and other Russian scientists; M. Fernex, a Swiss scientist; and C. Busby, a UK scientist, all presented reports on the very serious disastrous health effects caused by the accident at the Kiev conference. Yet, all these presentations were objected vociferously and suppressed not by scientific arguments but simply by denying such findings by the authorities, including S. Yarmonenko, the department of health of Belarus, and A. J. Gonzales, a representative of the IAEA, and N. Gentner of the UNSCEAR. They could not answer reasonably many questions posed by many Russian physicians. None of these findings of adverse health effects caused by the radiation was included in the final report from the conference.

It was also revealed that Y. I. Bandazhevsky, the president of Gomel medical school, whose very important medical findings about Cs-137 radiation are outlined in Sect. 14.3.2, was wrongfully accused to have accepted a bribe and jailed for 8 years. He was released because the allegation could not be proven, and was also aided by international concerned voices.

A movement called "ETHOS" was introduced by members of the Nuclear Protection and Evaluation Center (CEPN, associated with the ICRP). The main goal was to create conditions for inhabitants of the contaminated territories to enable them to become more autonomous actors in a rehabilitation process, embracing the improvement of the local living conditions, as well as increasing radiological safety (this is a quote from a brief description of ETHOS (2007)). No attention was to be given to health problems. It was implemented in 1996–1998 in a village called Olmany, Belarus; the reason that Olmany was chosen was because it was located in a zone where people had the choice to either stay in the area or move away, and the move would be financially assisted by the government. The area had Cs-137 contamination of 185–555 kBq/m^2, which corresponds to 1–5 mSv/year in terms of doses that individuals would receive. Six working groups were set up involving the villagers, and each group dealt with a specific task, such as the protection of children, radiological quality of milk production, etc. According to the organizer's account, the project was successful, in such aspects as improvement of living conditions regarding the private agricultural production and the radiological protection of children. The villagers gained a more precise and reliable picture of the radiological situation (Lochard 2007).

However, the pediatrician of the village reported disastrous results on people's health during the project in an international seminar (2001) on the ETHOS project. She mentioned declining health of the newborns, sudden increase of serious diseases among the villagers, only 20 % of children were healthy as compared to 80 % before the project, and that ten times more children were hospitalized compared to the period 1986–1987 in Blest (about 400 km west of Olmany). Her report was transmitted to the outside world by M. Fernex (Switzerland), who attended the seminar, but it was omitted from the official records of the seminar.

The ETHOS movement has no interest in the health of people, but is concerned with the economic issue of how to reduce the cost of dealing with the aftermath of the nuclear facility accident. In the same spirit, ETHOS was expanded to the Cooperation for Rehabilitation (CORE) project, which is still being operated in Belarus, under the support of the United Nations Development Program (UNDP: http://un.by/en/undp/db/00011742.html).

18.6 Fukushima Incident

The same strategies as outlined above are being used by the Japanese government, the nuclear industry, and the IAEA, ICRP, WHO, and others on the aftermath of the Fukushima Dai-ichi NPP accident. The only difference specific to Japan is the Japanese experience of atomic bombs. This acts in both ways: to favor or, rather, be ambivalent toward nuclear industry in some segments or to deny it.

The devastating effects of the atomic bombs are well remembered by the majority of the Japanese, probably better than the people of any other nation. However, the long-lasting adverse health effects on the atomic bomb survivors are not recognized very well. The government still denies the causality between ill health and the atomic bombs for many of the victims. The authoritative figure is based on the long-standing studies on survivors by the RERF, which ignores the internal exposure effects as detailed in Chap. 13.

The use of the atom for "military" purposes is hated and feared by the majority of the Japanese, but this has led, ironically, to an attitude favorable for the "peaceful" use of the atom. The sentiment among some atomic bomb survivors was that we were the victims of the military use of nuclear power, and, hence, we would like to see the useful aspects of nuclear power for "peaceful" purposes. So, we should stand for the peaceful use of nuclear power to compensate our sufferings. This attitude still seems to persist somewhat, and the main reason for it is the sense that "nuclear weapons" and "nuclear power reactors" are two completely different things; the former is evil but the latter is good. The fact that nuclear weapons and nuclear power reactors both produce a lot of radioactive material, along with a lot of energy, has not been recognized by many people. The evilness of the "nuclear" aspect is recognized only in terms of the uncontrolled release of the huge amount of energy (in the military application), but the common thread, i.e., radiation, is also "evil". This is, indeed, the theme of this book.

The most recent world congress of the International Physicians for the Prevention of Nuclear War (IPPNW) was held in Hiroshima in November 2012. The delegates from many countries wanted to include the issue of Fukushima, i.e., the radiation issue related to nuclear power plants, in its agenda. However, the Japanese organizers (the JPPNW, centered in Hiroshima) refused to do so; their decision represented the sentiment of some atomic bomb survivors, mentioned above, and, also, the Japanese government.

However, the majority of the Japanese are not in favor of the continued use of nuclear power as a part of their source of energy, recognizing a likely possibility of severe adverse effects on people's health due to the enormous amount of radioactive material released by the accident at the Fukushima Dai-ichi NPP. They no longer believe in absolute safety in the operation of nuclear power facilities, and, besides, Japan is highly vulnerable to earthquakes, with another strong earthquake being predicted to take place in the next 30 years or so.

Yet, the government, the nuclear industry, and the associated "experts" (scientists, medical doctors), in addition to the IAEA and ICRP (and the WHO), are all desperate to keep the industry going. They are now spreading a false sense of security that the radiation resulting from the Fukushima accident is low enough that no significant radiation effects will show up. They are trying to set the allowable radiation level high, even higher than the ICRP recommendation. However, the emergence of thyroid cancers (twelve, likely twenty seven cases in 24 months) among children in Fukushima is an undeniable sign of the seriousness of radiation effects on human health. The mass media dependent on the nuclear/electrical industry are silent about the seriousness, and they are not trying to look into numerous existing incidences of possible radiation effects (such as sudden heart failure, extreme fatigue, hair loss, etc.).

The Japanese government let the IAEA control the aftermath of the Fukushima accident. The ICRP has already begun a movement called "ETHOS", in which they allure people to deny themselves the seriousness of the situation and give a false sense of security and self-control, so that people would willingly choose to live in a contaminated area. It is supposed to be a movement by the people and for the people, so that the people would have no one to blame even if they faced a serious consequence, because it would have been their choice. However, it may be self-motivated, not inspired by the ICRP's ETHOS movement, and its intention to protect the people who stay in contaminated areas may be sincere.

Adverse health effects due to the radiation from the accident are slowly emerging, as mentioned in Chap. 15. Twelve cases (plus fifteen suspected cases) of thyroid cancer among Fukushima children have already been reported, for example. Yet, the official assessment of the health effect of the accident by the UNSCEAR as of December 12, 2012 (http://www.world-nuclear-news.org/RS_UN_approves_radiation_advice_1012121.html) included preliminary findings which showed that no radiation health effects have been observed in Japan among the public, workers, or children in the area of the Fukushima Dai-ichi NPP.

Incidences of adverse health effects seem to be increasing rapidly at the time of writing (March 2013), 2 years after the accident. Thyroid anomalies among children are now widespread and observed all over Japan. Sudden heart failure seems to be increasing among the young, too. Chronic fatigue syndrome is also becoming prevalent. All kinds of infectious diseases are increasing, very likely due to weakening of the immune systems in many people (see, for example, http://sos311karte.blogspot.ca/, English version).

Unfortunate trends are: (1) the government and the associated structure are in the mode of denial of adverse health effects of radiation; (2) the majority of medical doctors are not familiar with the radiation effects; (3) the major mass media are not keen to look into this issue (adverse health effects of radiation); and (4) they are influenced by the authoritative figures' insistence that only thyroid cancer is caused by radiation and all other effects have nothing to do with radiation.

References

AEC et al (1950) The effects of atomic weapons. USGPO, Washington, DC

Andreoli E, Cavazoni R, Tchertkoff W (2004) http://www.youtube.com/watch?v=0iyOisOy6LI

Bush-Conant file relating to the development of the atomic bomb, 1940–1945. Record of Scientific Research and Development, Record group 227, National Archives Microfilm Publications Pamphlet Describing M1392

Cooke S (2009) In mortal hands: a cautionary history of the nuclear age. Bloomsbury, New York

ETHOS (2007) Rehabilitation of living conditions in the territories contaminated by the Chernobyl accident: the ETHOS project in Belarus. ftp://ftp.cordis.europa.eu/pub/fp5-euratom/docs/ethos.pdf

Forshier S (2009) Essentials of radiation biology and protection, 2nd edn. Delmar, Clifton Park

Lapp RE (1962, 1963) Nevada test fallout and radioiodine in milk. Science, 137 (1962):756–758; Nevada test fallout. Science, 142 (1963):448

Lochard J (2007) Rehabilitation of living conditions in territories contaminated by the Chernobyl accident: the ETHOS project. Health Phys 93:522–526
Pflugbeil S, Claussen A, Schmitz-Feuerhake I (2011) Health effects of chernobyl. IPPNW & Gesellshaft für Strahlenschutz, Germany
Sawada S (2011) Estimation of residual nuclear radiation effects on survivors of Hiroshima atomic bombing, from incidence of the *acute radiation disease. Social Med Res 29:1–15
Sternglass E (1972/1981) Secret fallout: low level radiation from Hiroshima to Three Mile Island. McGraw-Hill, New York
Stewart A, Webb J, Hewitt D (1958) A survey of childhood malignancies. Brit Med J 1:1495–1508
Takahashi H (2008) Classified Hiroshima and Nagasaki: US nuclear test and civil defense program (in Japanese). Gaifu-sha, Tokyo
Takahashi H (2013) Investigation of radiation effects on human body during the Cold War period. J Jpn Sci 48:6–13

Chapter 19
The Crisis of Human Civilization and Life-Supporting Planet Earth

19.1 Nuclear Industry Lobby and Their Tactics

It has been shown in this treatise that radiation is, indeed, incompatible with life. Continued production and eventual release of radioactive material through military and "peaceful" uses of uranium and others is increasingly endangering the LIFE on this planet.

Unfortunately, this crisis is not recognized as such by the majority of the human race, particularly those scientists, politicians, and businessmen involved in the nuclear industry. Humankind has been enthralled and proud of their accomplishments in unraveling the secrets of the universe, with the atomic and nuclear structure research starting around the mid-nineteenth century. Indeed it is a great achievement, though far from complete.

The majority of the applications of science to solve some difficulties that humankind has been facing have been beneficial to humankind. An unfortunate tendency of humankind, though, is to apply scientific discoveries not necessarily for its "benefit". Well, many of the applications have not been for the "benefit for people", but, rather, for making "killing people more efficient" and, in the process, making corporations richer. This applies to not only the situation in the twentieth and twenty-first centuries, but also throughout the history of *Homo sapiens*.

A typical and most serious example is the development of nuclear weapons. The initial project was a secret national project. The government provided an enormous amount of money from tax collected from the people. The militarists, scientists, and corporations alike enjoyed the financial benefits of their work. Eventually, the work extended to nonmilitary, i.e., the so-called "peaceful" use of nuclear energy. Nuclear power generators are huge and complicated technical edifices, and building one requires an enormous amount of money. The major portion of this money was initially supplied by the government to enhance the development of the nuclear industry. All those involved are often called "Nuclear Lobby", "Residents of Nuclear Village", or other such names. They now heavily rely on the nuclear industry and

E. Ochiai, *Hiroshima to Fukushima*, Science Policy Reports,
DOI 10.1007/978-3-642-38727-2_19, © Springer-Verlag Berlin Heidelberg 2014

are desperately trying to defend the industry for the sake of their own lucrative lives, at the expense of the lives of others.

The main theme is that "low-level radiation" (often quoted as "below 100 mSv") is not harmful, and, so, people who live in contaminated areas do not need to worry about the effects of such a low level of radiation. The international regulatory commission (the ICRP) has been the major player in determining the allowable level of radiation. They assert that the regulation level has been determined scientifically, and many nations' regulatory bodies follow its recommendations. Their recommended lowest level has gradually increased over the years. It said in 1954 that the lowest allowable level should be: "to the lowest possible level", which changed to " as low as practicable" in 1956, then to "as low as readily achievable" in 1965, and it recommended levels "as low as reasonably achievable" in 1973. Their contention is that the safety level should be determined based on the cost–benefit considerations involved in living in a contaminated area. If the regulation level is set too low (i.e., too strict), the protection process is too costly, unnecessarily. In other words, the lowest level recommended by this body is determined not by science, but is decided based upon political and economic considerations. In this scheme, "life" and "living a safe and healthy life" are very much undervalued.

One disturbing strategy by the nuclear industry has been initiated by the CEPN/ICRP and tested on the people in connection with the Chernobyl incident, and seems to be applied now to Fukushima prefecture in Japan. It is called the "ETHOS" movement, which has been described earlier. The ETHOS movement in Fukushima may not be induced by the CEPN/ICRP, however. It is intended to reduce the cost of dealing with moving the people from a contaminated place to a cleaner place. That is, it induces the people to remain in a contaminated area, by convincing them and themselves that the radiation is under the control of the people and its effect on health is insignificant.

19.2 Other Signs of Radiation Effects?

In addition to the adverse health effects, other disturbing phenomena have been recognized that seem to have shown sudden changes in the last half a century or so. The first such data set is shown in Fig. 19.1 (Herrnstein and Murray 1994). It shows the change of the overall mean verbal scores of the US College Board SAT test. Starting around 1964, it plunged over the next 20 years. This had been predicted by Sternglass (1972/1981), based on his studies on the relationship between fallout and its health effects. The authors Herrnstein and Murray, however, attributed this change to the fact that minority young (black) had begun to enter colleges, and, hence, the change was due to the intellectual capacity or preparedness of the different ethnic groups. Other reasons suggested included changes in life-style, e.g., increase in TV viewing time. If, indeed, these are the main reasons, then the reversal of the trend starting in the early 1980s is hard to explain.

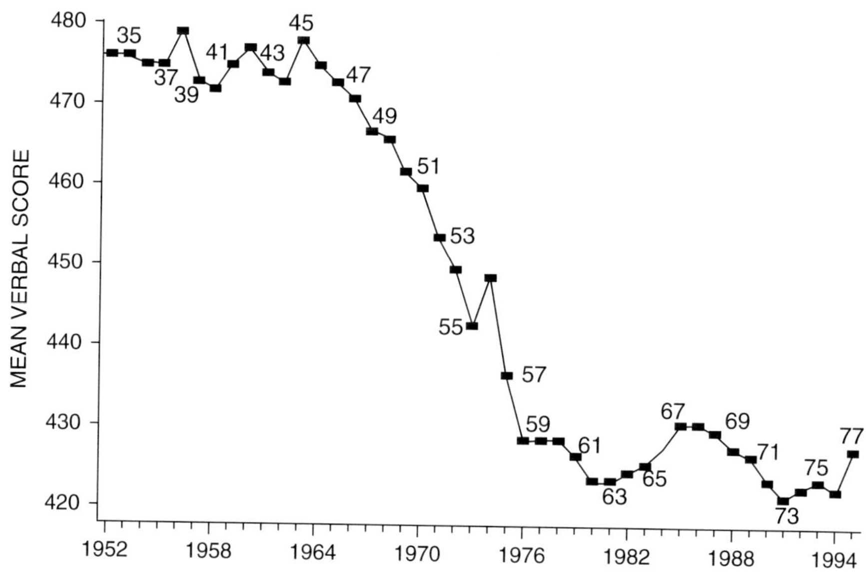

Fig. 19.1 US College Board SAT verbal scores (Fig. 3.1 in Gould 1996)

The numbers associated with the dots are the average year when the test takers were born. The decline started with the young who were born in 1946 (conceived in 1945) and flattened out and started to reverse with those born in 1964. This trend is parallel to the atomic bombs development: Trinity test, Hiroshima/Nagasaki, and then intense bomb tests in 1950–1960s, and the halting of major tests thanks to the US–USSR test ban treaty in 1963. It is consistent with the idea (Sternglass 1972/1981) that the fallout radiation from the atomic bomb explosions affected fetal development in the exposed women.

This effect has been manifested in the increase of leukemia and other cancers, and it is possible that exposure to low-level radiation at the fetal stage could affect the development of the brain. The regional differences of this score are also consistent with the regional variation of the likely fallout amount (Sternglass 1972/1981; Gould 1996). The general impact on intelligence of injuries in the brain by radiation is discussed in Sect. 14.5.3.

In 1996, a book titled "Our Stolen Future" (Colburn et al. 1996) startled many readers. One of the main focuses of this book was the "dramatic decrease of sperm counts" in many countries. A major paper was published by the Danish scientist Skakkebaek and coworkers (Carlsen et al. 1992). Along with the decrease in sperm count, cancers associated with the male sex organs increased, so it was reported. It has caused intense controversy and no definitive conclusion had been reached regarding the reality of the decrease in sperm counts. The cause for such a decline was attributed by the authors (Colburn et al. 1996) to the widespread dispersal to the environment of what is called "endocrine disruptors" such as diethyl stilbene (DES, which was widely prescribed for morning sickness and other ailments), dioxin, and polychlorobiphenyl (PCB). Many chemicals suspected of being endocrine disruptors have since been banned.

Another paper on the same topic was published recently (Rolland et al. 2013), and has awaken people to the fact that the issue is, indeed, real. This study involved more than 26,000 healthy French men (average age 35 years). It reports a significant and continuous 32 % decrease in sperm concentration over a period of 17 years (1989–2005). The annual decrease rate is about 2 %, and it is still decreasing. The quality of sperm also seems to be declining. The authors have not identified any causes for this phenomenon, citing mostly environmental effects, in particular, the endocrine disruptors.

It is well known that the male testicle is the organ which is the most sensitive to radiation, as mentioned earlier. Besides, the decline of sperm counts seems to be just opposite to the increase of the radiation on this planet, as shown in Chap. 16. These two facts, put together, may suggest that the sperm count decrease in the last 50 years or so may have something to do with the increase of the overall background radiation due to human activities. It is a complex issue and the cause may not be a single factor, and, indeed, the environmental effects and foods consumed (as suggested by some) are important, and environmental effects could encompass both endocrine disruptors and others, as well as background radiation. It may also have to do with the fact that the human population has been exploding the last several decades.

The extinction of species on this planet has been accelerating in recent decades, and the speed of extinction this time is much higher than that which took place in the past. No detailed data are available, and the true rate of extinction is hard to determine, for not all living species have been recognized as such by us, humans. The rapid rise of the human population and our quest for more materials and living space have disturbed and destroyed habitats for many living species. Whether the increase of background radiation has anything to do with the acceleration of extinction is not known and may not be discoverable, but there is a possibility that it may be at least contributing somewhat.

19.3 Beyond Radiation Issues

The radiation problem is only one aspect of the illnesses of current human civilization, but it is the most serious one, due to its potential danger to destroy life itself. Besides, it is a significant manifestation of the general trend that human civilization is now in. *Homo sapiens*, as the single intelligent species on this planet, are usurping all the material on this planet as if it is all theirs. Particularly the people in the so-called advanced nations are now using resources, including energy, beyond what the carrying capacity and the natural renewal rate allow. This trend is exacerbated by the actions of a minority in humankind. They exploit all the other fellow human beings, as well as all the resources for the benefit of only themselves. They have deliberately developed such a system under the guise of democracy, in which the economic and political systems are subjugated under their control. As outlined in this treatise, they (the minority oligarchs) use all tactics to suppress the truth of radiation effects, and the economic and political ill intentions, and, hence, the majority of people are shielded from the truth.

Often, this kind of attitude against nuclear power (or any other "advanced" technology) is labeled as "antiscience", "antiprogress", etc. It is not against science itself, i.e., the study of nuclear fission reactions and others, but is against the use of such science for the destruction of humanity, life, and the earth.

Perhaps it is necessary to touch on one more issue before closing, which is the issue of "climate change". It has been established that the climate on this planet has, indeed, changed significantly in the last decade or so, and is still changing. The main cause for this climate change has been attributed to the greenhouse effect by chemicals introduced by human activities into the atmosphere. Such chemicals include carbon dioxide and methane, both of which have increased during the last several decades. The source of carbon dioxide is the burning of carbon-containing compounds, particularly coal, petroleum, and natural gas. Currently, electricity is produced mostly (more than 85 %) by burning these substances. In these processes, about 40–60 % of the heat produced is converted to electrical energy, and the rest goes back into the environment. The increase of greenhouse gases may or may not be the main cause, but it is contributing to climate change.

The production of electricity by nuclear reactors is touted by the industry as a green energy source, to be able to break the ongoing warming trend, because, the industry claims, a nuclear reactor does not produce carbon dioxide, the greenhouse gas. This is true in its limited sense. That is, the nuclear fission reaction, indeed, does not produce carbon dioxide while producing a lot of heat. However, as pointed out earlier, only about one-third of the heat produced is converted to electricity, and the rest (two-thirds) is released into the environment. That is, the nuclear reactor is an excellent environmental heater, better than the other electricity production processes in this sense. This fact seems to be often ignored in the argument of the greenness of nuclear power.

The use of energy involved in the entire process of nuclear power production is very significant; mining, ore processing, uranium enrichment, fuel rod production, building and operation of the nuclear facilities, fuel recycling, and disposal of the radiation waste. Estimation of the energy consumption in these processes is not easy, and, besides, the last item (disposal of waste) cannot yet be evaluated meaningfully. The Australian government commissioned a comprehensive study on the issue, and the report is summarized in a review article by Lenzen (2008). The conclusions were:

1. Every kWh_{el} of electricity generated by a typical nuclear reactor requires, on average, 0.2 kWh_{th} (range 0.1–0.3).
2. This translates into an average greenhouse gas emission of 65 g CO_2/kWh_{el} (range 10–130).
3. The CO_2 emission in typical fossil fuel electricity generation is 600–1,200 CO_2 g/kWh_{el}, 15–25 g CO_2/kWh_{el} for the wind turbines and hydroelectricity, and 90 g CO_2/kWh_{el} for solar photovoltaic power.

The greenhouse gas emission from a nuclear reactor is obviously much less than a fossil fuel electricity generator, but even less than solar energy, according to this estimate. If the energy required as shown in (1) is supplied entirely by fossil fuel electricity, it seems that the nuclear facility would emit 180 g CO_2/kWh_{el} (of nuclear energy) (0.2 × 900 (=average of 600–1,200)). The study focused only on CO_2 emission and took no account of the heat released into the environment. It is not possible for this writer to evaluate how accurate and realistic are the numbers which the authors have come up with. This issue needs to be examined further.

A report examined the overall relationship between electricity generation and health, involving the environmental emission of harmful particles and others from electricity power plants and its effects on the mortality and morbidity of the public (Markandya and Wilkinson 2007). They state that nuclear power has one of the lowest levels of greenhouse gas emissions per unit power production and one of the lowest levels of direct health effects. The negative health effects of nuclear power in this analysis seem to be underestimated, ignoring largely the negative health effects discussed in Part V of this book.

Based mostly on these two reports (Lenzen 2008; Markandya and Wilkinson 2007), Kharecha and Hansen came to the following conclusion: "Nuclear power prevents more deaths than it causes" in a recent report (Kharecha and Hansen 2013). This conclusion needs to be carefully evaluated in view of the findings detailed in Part V, because it seems that the negative effects of radiation on humans and other organisms are not fully taken into consideration in these analyses.

References

Carlsen E, Giwercman A, Keiding N, Skakkebaek NE (1992) Evidence for decreasing quality of semen during past 50 years. Br Med J 305:609–613

Colburn T, Dumanoski D, Peterson Myers J (1996) Our stolen future. Plume (Penguin), New York

Gould J (1996) The enemy within, the high cost of living near nuclear reactors. Four Walls Eight Windows, New York

Herrnstein RJ, Murray C (1994) The bell curve: intelligence and class structure in American life. The Free Press, New York

Kharecha PA, Hansen JE (2013) Prevented mortality and greenhouse gas emissions from historical and projected nuclear power. Environ Sci Tech, (Web) March 15

Lenzen M (2008) Life cycle energy and greenhouse gas emissions of nuclear energy: a review. Energ Convers Manage 49:2178–2199

Markandya A, Wilkinson P (2007) Electricity generation and health. Lancet 370:979–990

Rolland M, Le Moal J, Wagner V, Royére D, De Mouzon J (2013) Decline in semen concentration and morphology in a sample of 26,609 men close to general population between 1989 and 2005 in France. Hum Reprod 28:462–470

Sternglass E (1972/1981) Secret fallout: low level radiation from Hiroshima to Three Mile Island. McGraw-Hill, New York

Postscript

An overwhelming amount of data seems to have been accumulated to indicate the very negative effects of radiation on life. The messages from these data/reports/books are out there waiting to be discovered and spread to the entire human race. This treatise has skimmed only the surface of such a depth and width of human knowledge.

Many of the data/analyses published, however, have been accused of having a lack of scientific rigor. Scientific research should be objective, for sure, implying that no bias, political or otherwise, should affect the entire procedure: obtaining, analyzing, and interpreting data, and publishing. In the case of health effects of radiation, it has been very difficult to be scientific in all the senses above, and also because of the nature of the phenomenon, as detailed in the text. However, if data/findings/reports that may conflict with the interest of the industry are kept suppressed or denied on the basis of the lack of scientific rigor, as is the situation currently, the present and future generations of humankind are deprived of opportunities to learn about and be cautious against the problems associated with radiation. We have presented some of the important data in Part V that reveal the adverse effects of radiation.

Unfortunately, this kind of treatise provides only the number: how many people have suffered from such and such cancers and other diseases. It cannot tell how terrible their sufferings and how miserable their lives are and have been as a result, and so on. This is the limitation of science.

However, many of the scientists and physicians who have been involved in unraveling the horrors of radiation effects have, indeed, involved themselves in directly helping those people who are suffering or have suffered. This book owes entirely to those brave and persistent scientists and medical doctors for its contents. The author expresses a deep gratitude to those persons. Particular thanks are due to Dr. S. Hida, a medical doctor, an atomic bomb survivor himself, who has voiced his experience in helping atomic bomb victims throughout his life, and Dr. S. Sawada, a physicist and also an atomic bomb survivor himself, who has endeavored to correct the official dose values of survivors and solve some of the mystery of internal exposure among the survivors. Dr. Ernest J. Sternglass, a physicist,

E. Ochiai, *Hiroshima to Fukushima*, Science Policy Reports,
DOI 10.1007/978-3-642-38727-2,

discovered the horror of the fallout from atomic bomb tests and then the radioactive emission from nuclear reactors, and persistently tried to publicize the serious issues, despite the strong opposition and suppression. Dr. Y. I. Bandazhevsky unraveled the mechanisms of internal exposure due to radioactive cesium based on direct observations of contaminated victims. Drs. A. V. Yablokov, V. B. Nesterenko, and A. V. Nesterenko studied the radiation effects caused by the Chernobyl accident and gathered all the relevant information in the form of a book published as an annual of the New York Academy of Sciences. There are many other scientists and medical doctors who have contributed to unraveling the effects of radiation. The publication of Yablokov et al. (2009) contains an enormous amount of data collected by these innumerable scientists and physicians. Dr. Helen Caldicott has been voicing the dangers of radiation in a number of publications accessible to the public, particularly in "*Nuclear Power is Not the Answer*" (Caldicott 2011). Our gratitude is due to all these people.

Finally, let us remind ourselves of some wisdom of ancient people. For example, Dine, Navajo Nation, has the following creation myth: "The people were given a choice of two yellow powders. They chose the yellow dust of corn pollen, and were instructed to leave the other yellow powder in the soil and never to dig it up. If it were taken from the ground, they were told a great evil would come" (LaDuke 2009).

A uranium mine was developed at Ningyo-toge in Japan in 1955 but closed in 1968. There is an old legend in the area, which says: "Do not enter, do not dig, do not play with it. If you do, you will be cursed". The tailings from the mining are still plaguing the people of the area.

Another legend was told by an old seer Louis Ayah to the Dene people around the Great Bear Lake in northern Canada. Until his death in 1940, he repeatedly warned his people that the waters in the Great Bear Lake would turn foul yellow, and that the yellow poison would flow toward the village. He also said that there would be sickness and that people would go through hard times and there would be deaths. The Canadian government opened a uranium mine in the area in 1932, and recruited the village men to transport uranium. Out of 30 men who worked there, 18 have died from cancers and lung diseases in the past 30 years in a village of about 600 people (Nikiforuk 1998). The ancient people knew the dangers of disturbing a "uranium" mine. They did not know about "uranium" or "radiation", but knew from their long experience that something evil would come out from certain areas, which are known today to contain uranium.

We should not increase the amount of radiation on the earth any more by disturbing uranium mines or producing and releasing new artificial radioactive material into the environment. Very limited and useful applications such as medical, industrial (tracer), scientific, and others may be allowed under strict control, though these are not absolute necessities.

"Hiroshima" (広島)literally means "wide island" and "Fukushima" (福島)means "happy island". The title of this treatise thus implies the following: To make the wide island (Entire Planet) a "happy planet". The author would be delighted if this little book may help people look seriously into the BIOHAZARDS OF RADIATION, help enhance more open and rigorous scientific endeavors about them, and reconsider the continued or increased use of nuclear power, both military and "peaceful".

References

Caldicott H (2011) Nuclear power is not the answer. The New Press, New York

LaDuke W (2009) The impact of uranium mining on indigenous communities. Orion, Jan/Feb 2009 http://www.orionmagazine.org/index.php/articles/article/4248/

Nikiforuk A (1998) Uranium haunts a northern aboriginal village. Calgary Herald, 14 March 1998

Yablokov AV, Nestrenko VB, Nestrenko AV (2009) Chernobyl: consequences of the catastrophe for people and the environment. Ann NY Acad Sci 1181:E1–E39